For my favorite chemist, — who makes good chemistry between us. / Love, Philip
Christmas 1993

*Chemistry Imagined*

*Roald Hoffmann and Vivian Torrence*

With a Foreword by Carl Sagan and a Commentary by Lea Rosson DeLong

# *Chemistry Imagined*

## REFLECTIONS ON SCIENCE

Smithsonian Institution Press · Washington and London

Editor: Frances Kianka
Production Editor: Jack Kirshbaum
Designer: Al Carter

Library of Congress Cataloging-in-Publication Data
Hoffmann, Roald.
Chemistry imagined / Roald Hoffmann and Vivian Torrence ; with a
foreword by Carl Sagan and a commentary by Lea Rosson DeLong.
p.     cm.
Includes bibliographical references.
ISBN 1-56098-214-4
1. Chemistry.     I. Torrence, Vivian (Vivian F.)     II. Title.
QD39.H68   1993
540—dc20                                                    92-31996

British Library Cataloging-in-Publication data available
Printed in Hong Kong.
96     95     94     93     5     4     3     2     1

♾The paper used in this publication meets the minimum
requirements of the American National Standard for Permanence
of Paper for Printed Library Materials Z39.48-1984.

# ■ CONTENTS ■

6

# ■ FOREWORD ■

Except for the two simplest, hydrogen and helium, atoms are made in stars. A cascade of thermonuclear reactions builds hydrogen and helium up into ever larger and more complex atoms which are then spewed out into interstellar space as the star ages and dies. There they drift for ages, occasionally coming close enough to one another to make a bond. Then two or more atoms make a commitment to go through life together. These bonds are the business of chemistry. In an eon or two a maelstrom of self-gravitating interstellar matter gathers up solitary atoms, and those bonded with their fellows, and plunges them into a forming planetary system. Four and a half billion years ago, that is what happened in our neck of the galactic woods. Our warm and well-illuminated little world is one result. All the atoms on Earth (hydrogen and helium still excepted) derive from these distant and ancient interstellar events—the silicon in the rocks, the nitrogen in the air, the oxygen atoms in a mountain stream; the calcium in our bones, the potassium in our nerves, and the carbon and other atoms that in exquisite detail encode our genetic instructions and job orders for making a human being. We too are made of starstuff.

There is hardly an aspect of our lives that is not touched fundamentally by chemistry: electronics and computers; food and nutrition; depletion of the protective ozone layer; mining and metals; medicine and pharmaceuticals; every disease including AIDS and cancer, schizophrenia and manic-depressive syndrome; drugs, legal and illegal; toxic water; and much of what we call human nature. We are, at least in large part, the way we are because of the atoms and molecules that make us up, and how they interact. In a deep and fundamental way chemistry makes us *us*.

So knowing at least something about chemistry is a prerequisite for knowledgeably functioning in human society, especially in our highly technological world civilization. Political decisions are made every day in the capitals of the world based on knowledge of chemistry. How, especially in a democracy, can the citizens influence decisions that their lawmakers make if we understand almost nothing about chemistry? Chemistry is not a required subject in the American school curriculum and few students learn it. You can watch prime time television for years and never come across a few minutes of coherent discussion of the subject. This is unwise.

Roald Hoffmann has recognized this problem and has attempted to help remedy it. He was responsible for a twenty-six-part PBS television series designed to explain chemistry to high school and junior college students and has been active in writing popularizations of chemistry. His collaborator, Vivian Torrence, responds to the beauty, mystery, and utility of chemistry in a set of striking collages. The present book couples these perceptive images with sprightly, bite-sized, and comprehensible essays on the widest imaginable range of chemical knowledge. It's very much the sort of thing we need more of.

Carl Sagan

# ■ ACKNOWLEDGMENTS ■

Nature's layers of reality are a constant source of inspiration for my art. The science of our time is particularly vital; it is a symbolic search, using the powers of logic and intellect as the driving force to find what is real. It lays the groundwork for our vision of the world. How complex our vision of the world is; so many points of view! How puzzling! Is it a surprise that I would meet Roald Hoffmann? In 1986 we met and began our conversations. What a remarkable person, this scientist and writer with the special vision to make connections between science and art. Recognizing in my work the interest in science, he asked, "Did you ever think about doing something with chemistry?" I had already investigated alchemy, which was full of visual symbols and spiritual energy. Now a rich, new world of symbol, search, and abstract thought would open and a collaboration of great artistic importance would begin. To Roald I wish to express my deep gratitude for his understanding and his trust in my work and in me. I am grateful for the series of events that have led to the beginning and to the completion of this special body of work.

I am also grateful to the many people in chemistry and art who have seen the value of our collaboration. I especially wish to thank Lea Rosson DeLong for her interest in my work and for her essay. Committed to depth of intellectual and spiritual understanding, her historical and personal insights have given me the inspiration and energy to make a transition to new

9

work. I wish to express my appreciation to Elisabeth Kirsch and Douglas Drake, who have over many years supported my work and given much energy and enthusiasm to *Chemistry Imagined;* also thanks to the supportive staff at the Douglas Drake Gallery and Elizabeth Meryman at *The Sciences.*

The collages for *Chemistry Imagined* were finished over a three-year period and in four different locations. The very first collages were done at Cornell University in Ithaca. I wish to thank the Art Department and the Council of the Creative and Performing Arts for sponsoring my stay on campus, along with members of the chemistry faculty, staff, librarians, and colleagues in related fields. In California, friends have given me the ongoing encouragement and support necessary: Mary Askew, Carolyn Brennan, Bear Capron, Dorothy Darr, John and Lisa Harris, Alva Henderson, Edith Hornor, Katrine and Sebastian Kuhn, Mary LaPorte, Janet Lewis, Joan Merrill, Anna Murch, Betty Orme, Susannah and Matt Phillips, Clint and Marilyn Smith, Katherine Solomonson, Sue and Andy Streitwieser, and Al Young. For their artistic advice and encouragement, I wish to thank Alberta Cifolelli, Randy Long, Kathryn Reeves, James Rosen, David Shapiro, Ron Slowinski, Gaylord Torrence, and Kay Walking-stick. I wish to recognize the efforts of European friends supporting my research and work in the early stages: Jorge Calado, Lisbon; Odile Eisenstein, Orsay; Michelle Goupil, Paris; Pierre Laszlo, Paris; and Ruslan Minyaev, Rostov-on-Don. I wish to thank Elisabeth Vaupel and Otto Krätz, of the Deutsches Museum for their encouragement. Also in Munich, I wish to give special thanks to Notker Rösch for sharing his knowledge of chemistry and mathematics, as well as giving his support, encouragement, and friendship during the completion of the series. I am grateful always to my long-term supporters Neal Benezra, James Demetrion, Maria Makela, Evan and Naomi Maurer, Jessica Rowe, Philip Florig, my brother, and most of all my mother, Edna Florig.

Vivian Torrence

Many of the people mentioned above aided and encouraged me as well. Several of the essays and poems in this volume were first published in various magazines; the judgment and editorial assistance of Michelle Press, Sandra Ackerman, and Brian Hayes at *American Scientist* was especially important to me. I am grateful to my friends and readers for their continued assistance, among them Ted Benfey, Derek Davenport, Thomas Eisner, Irving Geis, Peter Gölitz, Peter Jesson, Shira Leibowitz, Max Perutz, Carl Sagan, A. Truman Schwartz, William Shirley, Harriet Smithline, Howard Simmons, Frank Westheimer, and the multinational collective that is my research group. Zahava D. Doering played a critical role as a matchmaker in bringing this project and the Smithsonian Institution Press together.

My wife, Eva, has provided great moral, physical, and emotional support during this project, and has served as a careful, critical reader as well.

Roald Hoffmann

The single collage that is *Chemistry Imagined* was assembled with the support of so many institutions and people. The President of Cornell University, Frank H. T. Rhodes, and the Provost, Robert Barker, provided the essential seed funding. The primary support for our work was generously granted by the National Science Foundation, through its Program for Materials Development, Research, and Informal Science Education in the Directorate for Science and Engineering Education. Such opinions as are expressed by us in *Chemistry Imagined* are ours and not necessarily those of the foundation. The Council of the Creative and Performing Arts of Cornell University provided a grant to Vivian Torrence that allowed us to work together in Ithaca. To these institutions we owe much—they made an intricate creative process possible.

*Chemistry Imagined* was exhibited at the Purdue University Galleries, the Fine Arts Gallery of Indiana University, Augusta College, the Des Moines Art Center, the New York Academy of Sciences, the Douglas Drake Gallery, the Beckman Center for the History of

Chemistry, the National Academy of Sciences, the National Institutes of Health, and the Herbert F. Johnson Museum of Art at Cornell University. The organizers of these exhibitions put in much work on our behalf, work that we appreciate deeply.

Robert Ubell and Barbara Sullivan's faith in and much work on behalf of this project was very important to us. It was not easy to administer a research grant that spanned the continent, with an excursion to Germany. Joyce Barrows, Micci Bogard, and Sharon Drake at Cornell made it possible. At the Smithsonian Institution Press we valued the cooperation and guidance of Peter Cannell and the copy editing of Frances Kianka.

Finally, it was in a special place, the Djerassi Foundation, set in the inspirational Santa Cruz mountains, that we met and began this collage. We owe much to this wonderful place, to Carl Djerassi, and to the staff of the SMIP ranch.

Vivian Torrence and Roald Hoffmann

# Chemistry Imagined

# ■ FROM THE BEGINNING ■

. . . chemistry was the art of making substances change, or watching their spontaneous transformations. Ice turned into water, water could be made to boil. Grape juice or sugar cane mash turned into alcohol, and if you didn't intervene, it turned again, into vinegar. The colorless fluid in the gland of a Mediterranean sea snail, when exposed to air and sunlight, turned yellow, green, and finally a purple that could dye a skein of wool and hold fast.

The change, yellow into purple, A into B, solid into liquid, was as fascinating as it was useful. The metaphor of change prompted people to join a philosophy to simple chemistry, thus creating alchemy. And the sheer utility of transforming the world for our own purposes—making soap from lye and oil, mixing concrete and letting it set, learning to stop fermentation, making bronze from copper and tin that have been smelted from their ores—led to an evolution of lore into practice, then into industry and science.

Today chemistry is the science of molecules and their transformations. Over a period of a few hundred years art changed into science (with a convenient mythology to obscure how much art there still is in it), and instead of substances, chemists think of molecules. Of the original definition, *transformation* remains—colorless into purple, dangerous to innocuous (or the reverse), raw into cooked, molecule A into molecule B. Chemistry is about change, it always was, and will be. The change may be hidden, in the steel cylinders of a refinery, in the unseen but critical uptake of nitrogen by a bacterium. The change may be mathematicized, in a model of an atmosphere under stress above Mexico City. But the change is there.

The ancient art of making soap, a fancy scanning tunneling microscope, and a molecule of our oxygen carrier, hemoglobin, figure in this book. Because Vivian Torrence and I tangle with the spirit of chemistry, the book moves across time, bridges the ages of substance and molecule. Can one do this without the deadening jargon of science? Of course, but perhaps a brief introduction, a cross between a glossary and a trail guide, may help. The following, then, is offered so that you're not flustered by the chemist/writer's occasional lapse into the dialect of his trade.

*Pure/Impure.* It's not only your breakfast cereal (read the ingredients list!) that's a mixture—everything is. At the parts per million level, the purest spring water or pharmaceutical is a downright frightening mixture. Purity is an ideal; all real substances are mixtures of compounds. There are at least thirty-nine molecules making up the aroma of fresh cocoa.

*Compound.* The word implies that it's composed of something else. But whatever common salt or sugar is made of (we'll get to that in a moment), it persists, as salt and sugar. A compound is a substance, not immutable, but around for a most variable time, often for ages, under ambient conditions on this beautiful earth. And while we watch it, or abuse it, the compound maintains its characteristic physical properties—color, solubility in water, electrical conductivity, a melting point. Pure as we can get it, the compound is the building block of chemistry. A recipe for making aspirin (or a nerve gas) can be written down so that it works in Baghdad, Recife, or Wilmington, Delaware. Sometimes the chemistry works too easily.

*Molecules.* Atoms and, more important, recurrent groupings of atoms called molecules are what we would see, at the microscopic level, if we looked inside a compound. The word "see" is here used just as much to describe hard-won indirect knowledge as actual vision, aided by microscopes. Most of our knowledge of the structure of matter is indirect, yet it is as sure as the certainty that 1 gram of cyanide will kill you. In fact we know how that lethal action of a trivially simple chemical compound, $CN^-$, just a carbon atom connected to a nitrogen atom, proceeds at the molecular level.

What *does* one see inside? It depends on the "state" of matter. There are three common states of matter: solid, liquid, gas.

*Solids.* In a solid one sometimes sees seemingly perfectly ordered arrays of atoms marching off to infinity, as shown below for sodium chloride, NaCl, table salt:[1]

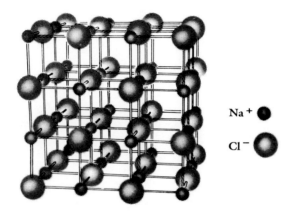

Na⁺ ●

Cl⁻ ●

Sometimes one sees ordered, repeating arrays of groups of atoms, as in a crystal of ice (here the larger circles are oxygen atoms, the smaller ones hydrogen; the dotted lines represent weak attractive forces called "hydrogen bonds"):

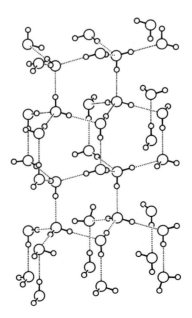

Or that of common sugar (technically, β-D-glucose):[2]

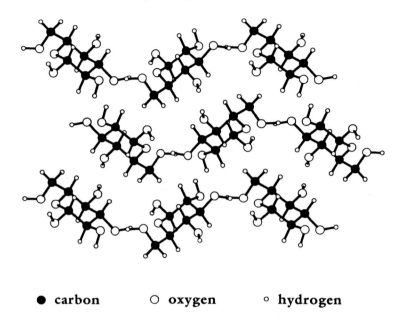

● **carbon**     ○ **oxygen**     ◦ **hydrogen**

Here we face for the first time complexity. And come to terms with it we must. Our bodies are not simple. In us, there are at least ten thousand chemical reactions going on. Difference, everything that makes the world exciting, derives not from simplicity but from complexity. That sugar is seemingly complicated (it really isn't) is intimately connected to the fact that it tastes sweet to us, but other molecules don't.

In the solid state the atoms or molecules are not sitting still. If we could observe them, and if we could look very quickly, we would see them moving, tethered to their sites, but moving. And the hotter it gets, the more they move.

*Liquids.* In the liquid "phase," as it is called, the molecules move off their fixed sites. They begin to move more quickly and randomly. They clump together for a while, then part. Still, their molecular identity is fixed. So in molten sugar or in liquid water, *our* liquid, one would see individual sugar or water molecules. Here's what a computer says a snapshot of liquid water looks like.[3] A thousandth of a second later all the dancers on this mad ballroom floor would have changed their positions.

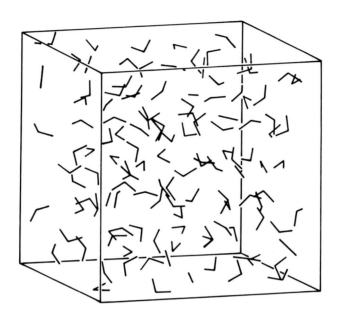

*Solutions* are a special case: mixtures that are liquid, two or more molecules swimming in the same space, weakly or strongly attracted to each other. It is easy to transport liquids, and on the molecular scale transport is easy within the liquid state. That is why solutions are universally used (in the body, in the laboratory) to carry out reactions. So many molecules dissolve in our favorite solvent, water. And for those that don't dissolve in $H_2O$ we have alternative solvents at hand—oil, alcohol.

*Gases.* In this simplest form of matter, the molecules' free motion overcomes the attractive forces between them. They are still sugar or water (steam, we now call it); not a single atom is lost. The molecules in a gas move quickly, collide often. Through their collisions they bang out an effective inviolable space, much bigger than they are—just like dancers.

If you insist on knocking a molecule apart into its component atoms, it can be done with heat or light, with the input of energy. But under normal conditions on Earth, the incredible multitude of matter is in molecular form. Water does not fall apart into one oxygen and two hydrogen atoms. It remains $H_2O$, whether it be ice, water, or steam. Near the sun it's different; molecules of water would not survive long as molecules there.

*Small* they are, these molecules. Which is why they were so difficult to "see," although their existence was presciently posited on philosophical grounds by the Greeks.

Here is the sugar (glucose) molecule drawn to scale. Just as a map might be 1:250,000, meaning that an inch on the map is 250,000 inches in reality, so we could say that these drawings are roughly 122,000,000:1, 1.22 inches on this page corresponding to 0.00000001 inches in the molecule. It is much too small to be seen with the eye, or even with the very best optical microscope.

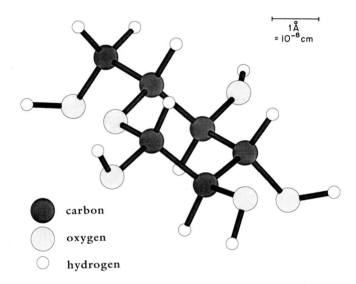

carbon

oxygen

hydrogen

*Atoms.* As we know well, these are the building blocks of molecules. So water has three atoms in it, glucose twenty-four.

We'd like atoms to look like this:

Indeed this is how they were represented in a model developed by Niels Bohr. The model had a short life span (1913–26), but it retains a strong hold on our imagination because it seems planetary, drawing on our hard-won astronomical experience. We *want* things inside to be the same as outside.

The model was replaced in 1926 by modern quantum mechanics. We now see the atom as still nuclear, in the sense of a small, massive, positive nucleus with negatively charged electrons around it. The electrons move in such a way that only the average features of their motion are available to us. And that seemingly incomplete knowledge suffices.

This is the electron cloud picture; the density of the cloud indicates the likelihood of the electron being there.

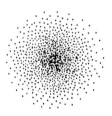

The planetary model is gone, but as a metaphor it remains ineradicable, absolutely ineradicable.

*Bonding.* Atoms are nice, atoms are fundamental, but they're not chemistry. Chemistry is about molecules, the fixed but transformable way in which atoms get together for a while.

Under normal conditions on this planet, water, in all its forms, is made up of $H_2O$ molecules:

What holds these atoms together, what is the glue that resists this molecule's falling apart into two H and one O atom? It is the chemical bond, and it happens to be what I make a living explaining.

As two atoms approach they begin to feel each other. The electrons of one molecule are attracted to the nuclei of the other, yet are repelled by the other atom's electrons. A balance takes shape. Sometimes it culminates by shifting an electron from one atom to another. This is called ionic bonding, as in salt. Sometimes no such shift occurs, but some electrons are shared by both atoms. This is covalent bonding, prevalent between the atoms of sugar.

The glue is strong. At normal temperatures the atoms cohere, move as a group. Perhaps a better analogy is that the chemical bond is like a spring that is crafted, naturally, between the atoms. It holds, but also gives, so that the atoms move a little around positions that are fixed on the average. These sites define the structure of the molecule.

*Molecular Structure.* In the gas or seething liquid, an individual molecule is buffeted by collisions. It travels, gives, deforms, yet remains a molecule. The structure that it struggles to retain endows all discussions of molecules with an architectural tone. Chemists, even as they know that molecules are not rigid, see them as static structures—at least some of the time.

The static perspective is a fruitful world view, for we are builders from the beginning. First, you build simple things, like a cube made of eight carbon atoms, each with a hydrogen pointing out. Devilishly hard to make, as it turned out—I know of dozens of people who tried but failed to make "cubane" before Phil Eaton and Tom Cole succeeded in 1964.

Or you can build more complicated things. Here is a new, effective, immunosuppressant, FK-506:

Then you can fiddle with the design. If you can find out how this immunosuppressant fits the site that it binds to in our body, perhaps you can change a piece of the drug that seems to produce unwanted side-effects.

Note how the architectonic language lends itself not only to statics but to dynamics as well. There is change in the very act of building. And the structure itself invites rebuilding, variation. Andrea Palladio's villas were all different. The building that chemists do is architectural, but it is a curious, remarkable kind of building. You put some molecules in a flask, input energy, and, hands-off, $10^{23}$ molecules do your bidding, sew up a bond here, break one there.

The blueprint for action on the molecular level is drawn in formulas, fleshed out with molecular models.

*Representing Molecules.*[4] Given that molecules are made of atoms and that the atoms are held together by bonds, how do we draw them? People outside of science, wishing for a center that holds, may think that science has it. Chemistry, or physics, captures reality and represents it faithfully. I'm sorry that I will disappoint you. There *are* glimpses of reality, to be sure; our tools, all those fancy spectroscopies, give us pictures through different, tinted glasses. What we represent (when we draw a molecule) is what we want to represent: we abstract a piece of reality to show it to another person.

Take camphor, a waxy medicinal substance with a characteristic, penetrating aroma. Here is the way most chemists are likely to see it in their professional journals, to fix it in their mind, to love the molecule:

1

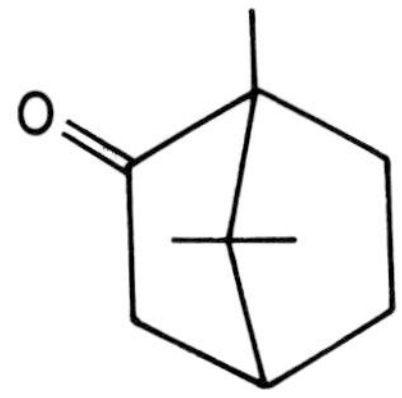

We know molecules are made of atoms, but what is one to make of the polygon of structure **1**? Only one familiar atomic symbol, O for oxygen, emerges.

Well, it's a shorthand. Just as the military man gets tired of saying Commander in Chief, Pacific, and writes CINCPAC, so the chemist tires of writing all those carbons and hydrogens, ubiquitous elements that they are, and draws the carbon skeleton. Every vertex that is not specifically labeled otherwise in structural representation **1** of camphor is carbon. Since the valence of carbon (the number of bonds it forms) is typically four, chemists privy to the code will know how many hydrogens to put at each carbon. The polygon drawn above is in fact a graphic shorthand for structure **2**. At several places in this book you will see such abbreviated structural formulas for carbon and hydrogen-containing molecules.

2

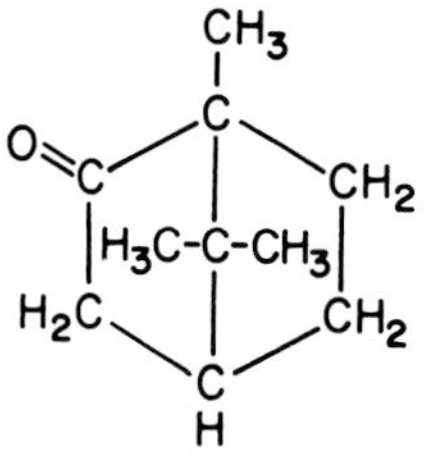

But is **2** the true structure of the molecule of camphor? Yes and no. At some level it is. At another level the chemist wants to see the three-dimensional picture, and so draws **3**:

**3**

At still another level, he or she wants to see the "real" interatomic distances, that is, the molecule drawn in its correct proportions. Such critical details are available, with a little money, a little work, by a technique called X-ray crystallography. And so we have drawing **4**, likely to be produced by a computer:[5]

**4**

This is a view of a so-called "ball-and-stick" model, perhaps the most familiar representation of a molecule in this century. The sizes of the balls representing the carbon, hydrogen, and oxygen atoms are somewhat arbitrary. A more "realistic" representation of the volume that the atoms actually take up is given by the "space-filling" mode! **5**:

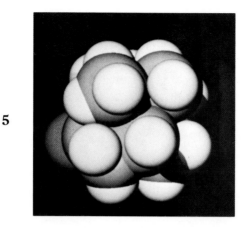

**5**

Note that in **5** the positions of the atoms, better said of their nuclei, become obscured. And neither **4** nor **5** is portable. It cannot be sketched by a chemist in the twenty seconds that a slide typically remains on the screen in the rapid-fire presentation of the new and intriguing by a visiting lecturer.

The ascending (descending?) ladder of complexity in representation hardly stops here. Along come the physical chemists to remind their organic colleagues that the atoms are not nailed down in space but are moving around those sites. The molecule vibrates; it doesn't have a static structure. Another chemist comes and says: "You've just drawn the positions of the nuclei, but chemistry is in the electrons. You should picture the chance of finding them at a certain place in space, the electronic distribution." One tries to do this in **6** and **7**.

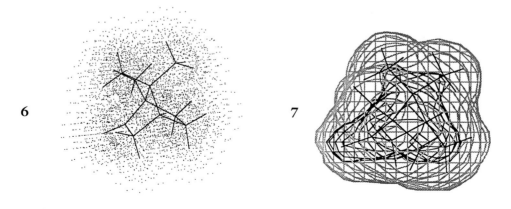

**6**

**7**

We could go on. The literature of chemistry does. But let's stop and ask: Which of these representations, **1** through **7**, is correct? Which *is* the molecule? Well, all of them are, or none are. Or, to be serious, all of them are models, representations suitable for some purposes, not for others. Sometimes just the name "camphor" will do. Sometimes the formula, $C_{10}H_{16}O$, suffices. Often it's the structure that's desired, and something like **1** or **3** is fine. At other times one requires **4** or **5**, or even **6** or **7**. Most of the chemical structures you will see in this book, and there are not many, will be drawn in a representation such as **2**.

*Chemical Reactions.* We return to that part of the definition of chemistry that has survived from medieval times to today: chemistry is change. While the atoms in a molecule persist in their association with each other, the input of energy—heat, light, electricity—can induce a change. From collisions on that busy dance floor emerge regroupings, new associations of atoms, new molecules.

Some transformations are slow, some are over in a twinkling of an eye. Building cement sets in hours, and it's easy to see why we wouldn't want it to set in seconds or years. But the cement in a dental restoration should set in minutes. And the skilled hand that puts that restoration into your mouth responds to the eyes' indication in milliseconds.

How do the eye, brain, muscle (and the cement) run their changes? We want to know.

What happens in chemical reactions may be described in many different ways. You can show a series of snapshots, as of this vigorous, to put it mildly, reaction of bromine with aluminum:[6]

Or you can describe the process in words:

Into a 300 milliliter beaker containing 10 grams of orange-brown liquid bromine (with some bromine gas evaporating above that liquid) I placed a few grams of aluminum. They reacted so vigorously that the aluminum melted, glowing white hot. At the end the beaker was coated with aluminum bromide.

Or write a chemical equation:

$$2\,Al + 3\,Br_2 \longrightarrow Al_2Br_6$$

All that is on the surface; it is what we can see. But how does it happen down there, on the microscopic level? What is the *mechanism* of the reaction, the sequence of elementary reactions that, taken together, make the process that we observe?

Sture Forsén, a Swedish chemist, has written:

The problem facing the scientist has been compared with that of a spectator of a drastically shortened version of a classical drama—"Hamlet" say—where he or she is only shown the opening scenes of the first act and that last scene of the finale. The main characters are introduced, then the curtain falls for change of scenery and as it rises again we see on the scene floor a considerable number of "dead" bodies and a few survivors. Not an easy task for the inexperienced to unravel what actually took place in between.[7]

Different kinds of snapshots, not necessarily photographic but using incredibly ingenious devices, show the sequence of elementary steps in a chemical reaction in exquisite detail. We now have ways of getting one of these snapshots in 0.000000000000001 seconds!

You can also freeze a reaction, slowing down its steps. You can throw in wrenches of different size and shape, molecular wrenches, to foul up this or that piece of molecular machinery. From what is then not made, or what is made misshapen, you can infer what makes what. You can send in atomic-sized spies to seduce the molecules from their appointed tasks. There are ways to learn the innards without seeing. Chemists are very good at that.

*Chemistry.* Compounds, atoms, molecules, bonding, structure, models, reactions. This is the conceptual stuff of the trade, much like the paint brushes, pigments, and canvas of the artist. You can describe in a narrative what the artist does, actually transforming these elements and tools into a watercolor. That description will lack something essential, namely, the spirit of a human creation. Vivian Torrence and I essay one collage-like representation of that spirit in these pages.

Roald Hoffmann

# ■ R A D I U M ■

In 1947 I was ten years old. We were in a DP ("displaced persons") camp in Wasseralfingen, then in the French Occupation Zone of postwar Germany, waiting for a visa to come to the United States. Or maybe we'd go to Israel. Or, in the desperate moments when the visa seemed unattainable, my stepfather even thought of signing a labor contract (in exchange for a visa) to work in the mines in Chile.

I was becoming proficient in my fourth language, German, and doing well in school, a school typical of the period, where every class had kids of different ages, for who had gone to school during the war? I read much, and somehow there came my way two books, biographies of scientists. One was of George Washington Carver, the black agricultural chemist, the other the biography of Marie Curie by her daughter Eve. I read both in German translation.

In the story of Carver I was fascinated by the transformations he wrought with the peanut and the sweet potato. Ink *and* coffee from peanuts, rubber and glue from the sweet potato! Perhaps part of the romance was that I had never seen or tasted either peanuts or sweet potatoes.

My Polish background certainly provided a ground of empathy for watching Manya Sklodowska transformed into Marie Curie. But Eve Curie's story touched something deeper. I remember to this day the scene when Pierre and Marie complete the painstaking isolation

*Radium* (1991)

of a tenth of a gram of radium from a ton of crude pitchblende. They put the children to bed and walked back to their laboratory. I must quote now, from Vincent Sheean's translation:

> Pierre put the key in the lock. The door squeaked, as it had squeaked thousands of times, and admitted them to their realm, to their dream.
>
> "Don't light the lamps!" Marie said in the darkness. Then she added with a little laugh:
>
> "Do you remember the day when you said to me 'I should like radium to have a beautiful color'?"
>
> The reality was more entrancing than the simple wish of long ago. Radium had something better than "a beautiful color": it was spontaneously luminous. And in the somber shed where, in the absence of cupboards, the precious particles in their tiny glass receivers were placed on tables or on shelves nailed to the wall, their phosphorescent bluish outlines gleamed, suspended in the night.
>
> "Look . . . Look!" the young woman murmured.
>
> She went forward cautiously, looked for and found a straw-bottomed chair. She sat down in the darkness and silence. Their two faces turned toward the pale glimmering, the mysterious sources of radiation, toward radium—their radium. Her body leaning forward, her head eager, Marie took up again the attitude which had been hers an hour earlier at the bedside of her sleeping child.
>
> Her companion's hand lightly touched her hair.
>
> She was to remember forever this evening of glowworms, this magic.[8]

Years have passed. The boy whose interest in science was stirred by German translations of a story of a black American applied scientist and a French-Polish woman chemist, is older. He rereads these books, and sees that they are hagiographies. The romance is off the radium. But Marie Curie still makes him cry.

There is only one great poet I know who usually does not need a second draft. Most of us revise, as William Blake did, for his well-known poem *The Tyger,* from "Songs of Innocence" (below, *left*):[9]

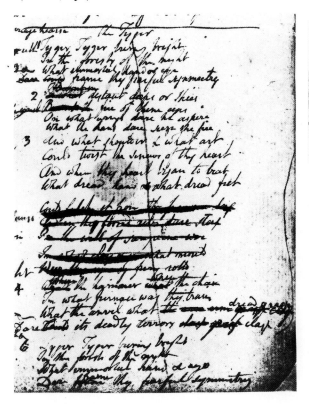
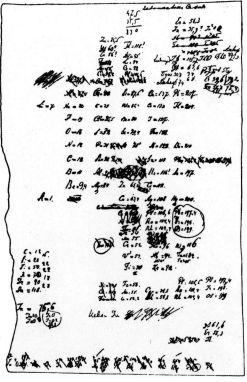

It's a great source of comfort for those of us who must change, to improve, or ruin, and change again, to see Blake struggle for the right word, to cross out "the arms," replace it by "grasp," by "clasp," by "dread grasp."

Seventy-five years after Blake, Dmitri Ivanovich Mendeleev puts the atomic weights of the elements on cards. He lays them out, reshuffles, and arranges them in different ways. It's a solitary game, this atomic patience. The emergent pattern has to be written down for a paper Mendeleev will present; pictures of cards will not do, so he draws a table (above, *right*).[10]

This is a draft too, as much as Blake's is. Titanium next to silicon is crossed out, hydrogen is moved. At the bottom is a tally of elements to be fitted into the table. Above it, in abbreviated Russian, it says "Don't Fit: In, Er, Th, Y." He makes them fit. The draft bespeaks the act of creation, that that act is human. We can all aspire to it.

Mendeleev let the numbers (here the atomic weights of the elements) speak, as Blake did, delineating in another language, another time, the wonder of the tiger. Mendeleev did not let the numbers tyrannize him. The atomic weight of tellurium was thought in his day to be greater than that of iodine. But he had reasons—the patience pattern for placing tellurium before iodine—for postulating that the experimental atomic weight determinations were slightly incorrect. He had the courage to leave blanks in his table, the elements that skillful hands had yet to discover.

The "fearful symmetry" of the elements was indeed framed by a Russian scientist. Mendeleev conceived this incredibly useful icon of chemistry without any understanding of *why* the Table was the way it was. In his own words: "It has been evolved independently of any conception as to the nature of the elements. It does not in the least originate in the idea of a unique matter and it has no historical connection with that relic of the torments of classical thought." Mendeleev was alluding to Platonism—he despised classical philosophy (later he said that what Russia needed more than one Plato was two Newtons).

For understanding one had to wait forty-four more years, six years after Mendeleev died, until 1913, when Henry Moseley and Niels Bohr explained the shell structure of the atom. How much good chemistry would have been lost if one had waited, immobile, for that understanding! As much as would have been lost in some fancied reductionist universe where the composition of poems prior to total scientific understanding were forbidden.

*The Periodic Table* (1991)

# ■ AMAZING GROWTH ■

Here's the problem: Nearly every molecule in our body contains nitrogen (N). N is there in amino acids, the building blocks of proteins, in our genetic material, in molecules that transport energy inside the body. Where do we or animals, or the plants that we consume, get our nitrogen from? Well, there is all that nitrogen around: some of it in mineral deposits, but especially prominent is the vast storehouse of the atmosphere, some 78% of it diatomic $N_2$ molecules. We breathe the air, take in all that nitrogen. We use the oxygen part of the atmosphere; could it be that we, such clever end-products of billions of years of evolution, would not use the equally available nitrogen?

The answer is that we most certainly do *not* use atmospheric nitrogen directly. We inhale it, and we promptly exhale it. It even dissolves in our blood, quite a bit of it. It goes for a merry ride around and comes right out, not having helped us, not having harmed us. Unless we are in the high-pressure environment of a diver. They have to worry about nitrogen narcosis, similar to alcohol intoxication, and the bends, when too rapid a relief of pressure causes bubbles of $N_2$ to form in the blood.

We get our nitrogen from plants. They in turn get it from two main sources: ammonia ($NH_3$) and nitrate ($NO_3^-$) from the soil or microorganisms. Higher plants also cannot "fix" the nitrogen of the atmosphere; their nitrogen originates from natural sources and chemical

*Amazing Growth* (1991)

fertilizers. And there is a remarkable chemical and human story behind each of these.

Some of the natural sources are minerals, decomposing plant and animal life, and an unexpectedly massive amount of nitrate produced from atmospheric oxygen and nitrogen by lightning discharges. The nitrogen comes down in nature's own, absolutely essential, acid rain.

But the most remarkable nitrogen fixer is a class of bacteria, living in nodules on the roots of leguminous plants in a symbiotic relationship with them. Legumes can fix tremendous quantities of gaseous nitrogen, or rather the bacteria coexisting with them can.

It's nice, I think, that one has to go so far down (or back) on the scale of biological complexity to find organisms that provide a necessary nutrient for nearly *all* more complex organisms. The adaptive advantages of specialization and symbiosis are striking.

How do the bacteria do it? They have (and we don't) an enzyme, a small chemical factory, that takes $N_2$ to $NH_3$. This enzyme is nitrogenase, prosaically named after the task that it does so well. And it has a surprise in its innards.

Enzymes are proteins, long assemblies of slightly differing amino acids. Small variations in the building blocks provide the opportunity, and the chain-like ganging of the amino acids provides the necessary diversity, to guarantee specific geometry and function. There is usually an active site in any enzyme, where the actual chemical task (here taking atmospheric nitrogen to ammonia) is accomplished.

What is different about nitrogenase is that in addition to the normal amino acids of all enzymes, this one has near the active site a cluster of sulfur and two kinds of metal atoms: iron and molybdenum. Iron—well, we've gotten used to that element in life—it's in hemoglobin and in other critical biological molecules. But molybdenum? What is molybdenum doing in a biological system? And such an important one at that! For without those molybdenum atoms no nitrogen will be fixed by bacteria.

The correct response is first: I'm glad you asked. That's the fun of chemistry, finding seemingly stodgy, seemingly rare inorganic metal atoms lurking inside a critical biochemical mechanism. Second: It's probably an accident, a billion years old. No plan, just something that worked.

In what form is that molybdenum and iron? How does it do so efficiently what all the fancy enzymes in our bodies can't do? Until 1992 we didn't have an inkling of an answer. Not for want of trying. An intense international competition has been on for the structure of nitrogenase. I know groups in Russia, China, England, and the United States that were and are in the race.

Other scientists are focusing on the Mo-Fe-S cluster at the active site of the monster protein—the whole beast weighs about 8,500 times the nitrogen molecule it processes. The cluster-builders are synthesizing molecules of iron, molybdenum, and sulfur in the laboratory, hoping that the natural activity of nitrogenase could be modeled by molecules much smaller than the protein. Here is a selection of the attractive polyhedral clusters they have made:

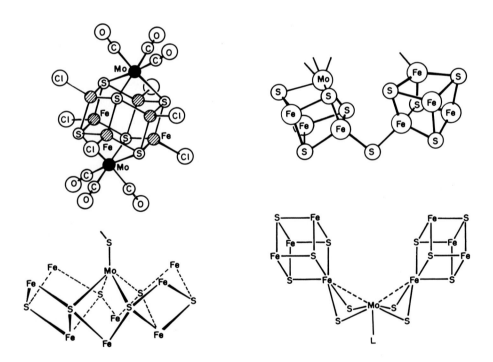

Some came close, but none turned out to be correct. In the summer of 1992, Douglas Rees and his coworkers at Caltech, after a decade of work, proposed a structure for the active Mo-Fe-S cluster, based on X-ray diffraction measurements. Here it is:

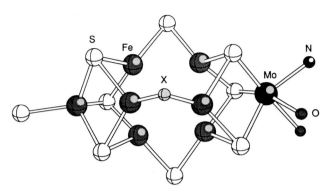

The active site indeed consists of a polyhedral array of seven irons, one molybdenum, and some sulfurs. The elongated molecule has a place where the nitrogen about to be "reduced" to ammonia might dock. It's where the bridging X sits (X indicating a group of atoms containing N or O; the picture X-ray diffraction gives is not yet accurate). Richard Holm, an active worker in the field, has called the Rees study "the protein structure of the decade."

Let's return to synthetic fertilizers, the mainstay of modern agriculture, the other main source of nitrogen. Urea and ammonium nitrate are the primary fertilizers—an easily stored and applied slurry of these with water goes under the name "liquid nitrogen." Both are made from ammonia, the no. 5 item on the world's chemical production hit chart in 1991. The commercial procedure for making much of the ammonia is called the Haber-Bosch process.

Fritz Haber, the man who invented the modern ammonia synthesis, was born in 1868 in a German-Jewish family in Silesia. A poem elsewhere in this book tells of his complicated life. He was unusual—combining a respect for the deep knowledge of the fundamentals of chemistry with a practical bent, a willingness to attack complicated industrial problems. The process Haber ingeniously developed around 1910 took atmospheric nitrogen and hydrogen gas (a product of processing natural gas) to ammonia at high pressure (250 atmospheres) and relatively high temperature (around 400° C), with an iron catalyst. Without the catalyst the reaction goes too slowly.

Perhaps the story here is not so much about nitrogen as about catalysts—agents that become involved in a reaction, speeding it up, agents that then are regenerated, unchanged, to run the reaction again. Nitrogenase, with molybdenum at its core, is a catalyst. So is Haber's iron. But Haber was not a catalyst; no human being can be. He was involved, in his tragic way. He was consumed.

# ■ ENERGY AND FORM ■

How can something as simple and homogeneous as the air be a mixture of several gases? And in that gentle, rarified fluid could there possibly be a seething mass of molecules in motion, tiny things propelled at a speed exceeding that of a Boeing 747, but traveling only a billionth of a second before colliding with another?

You can see the problem for both the chemist and the physicist who slowly extracted a confession of its consistency and properties from recalcitrant thin air. The problem persists even for us today, as we try to reconcile that reliable knowledge with daily experience.

That a simple, clear-looking substance could be a mixture is obvious. Take water, dissolve in it sugar, salt, baking soda, an antacid (read the ingredients, they may be complex), even worse. It still looks like clear water, but you know it isn't pure. Similarly, a sweet disposition, or intelligence, or kindness to children, is in and of itself not a guarantee of moral good in a person.

As for chemistry in the air—well, you know what happens when you bring a match up to natural gas streaming from your gas stove. Or take a look at the air above Los Angeles or Mexico City on two different days. There's a lot of chemistry going on in and with air.

Air is a potentially reactive mixture of some 21% oxygen, 78% nitrogen, 1% argon, and tiny bits of other gases: carbon dioxide, water, sulfur and nitrogen oxides, methane, ozone, neon, argon, xenon . . . We live on the oxygen, while other partners in the biosphere (some species of bacteria) "fix" the nitrogen of the atmosphere for plant and animal use.

Ozone, so much in the news lately, instructs us on the energies and forms of the air. This $O_3$ molecule is another allotrope, or alternative form, of the element oxygen. The normal form of the molecule we breathe is diatomic, $O_2$. Another, unstable form of the element is atomic oxygen, just O. Ozone by itself is a pale blue gas with a pungent odor that we associate with electrical discharges in air, which in fact produce ozone. The gas is very reactive, indeed corrosive. At sea level it is a primary component of photochemical smog, damaging tires, paint, and human tissue.

But there are no molecular villains, only molecules reacting with other molecules. That nasty actor at sea level, ozone, forms a dispersed layer in the stratosphere, concentrating about 25 kilometers up above the earth's surface. There it protects us from the most dangerous component of the sun's ultraviolet radiation by simply absorbing that high-energy light. There is never much ozone in that sunscreen; at sea level all the ozone in the atmosphere would form a layer only 3 mm thick.

Now, it's not just sitting there, the ozone. It's formed by the reaction

$$O_2 + O \rightleftharpoons O_3$$

The reverse of that reaction is also a way that ozone decomposes, which is why the above chemical reaction is written with two arrows. There are two other ways in which ozone disappears:

$$O_3 + O \longrightarrow 2O_2$$
$$O_3 \xrightarrow{h\nu} O + O_2$$

Here $h\nu$ is the symbol for the energy of sunlight absorbed.

Where do the O atoms come from? Also from photochemistry:

*Energy and Form* (1989)

$$O_2 \xrightarrow{h\nu} 2O$$

If that were all, there would be much more ozone than there is up there. The concentration of ozone is kept down, naturally, by a set of reactions involving catalyst molecules X:

$$X + O_3 \longrightarrow OX + O_2$$
$$OX + O \longrightarrow O_2 + X$$
$$X = OH \text{ or } NO \text{ or } Cl$$

OX is a molecule that is formed and disappears. It's called an "intermediate." X gets involved in the reaction, but is reformed. This phoenix is a catalyst. The net reaction is $O_3 + O = 2O_2$, that is, more of one of the reactions written out above. The natural controls on ozone are through catalysts X = OH, NO, and a little bit of Cl (chlorine). These molecules are not stable at sea level but are themselves produced in the atmosphere from entirely natural precursors. For instance, OH comes from water, HOH.

What have *we* done? We've introduced into the atmosphere a steady man-made supply of X = Cl, much more than was there before. The chlorine atoms, Cl, are formed by sunlight decomposing chlorofluorocarbons. These Cl's attack the ozone layer, diminishing it.

The atmosphere, seemingly featureless, inert, motionless, harbors shapes within, such as ozone, Cl, and OCl. And it selectively transmits and responds to energies, those of the sun and ours. The air, thin as it might be up there, is in contact with the earth underneath, and the sun above, through the magnificent cycles of matter and energy.

# ■ G R E E K   A I R ■

When Jan Baptista van Helmont (1577–1644), a Dutch scientist and alchemist, first introduced the word for a gas, he presciently derived it from the Greek word *chaos* (perhaps tran-
scribed roughly into Dutch). In Greek mythology this term described the unformed mass of
the universe before the creation of the gods. The Greek word has also come down directly
into English, in the primary sense of a state of confusion and disorder.

This is precisely how gases are, though it took 250 years after van Helmont to reach this
understanding. First, it should be said that English words, the words of any language, are
never divorced from their associated meanings (which is what makes our language so rich).
There is no negative connotation attached to disorder in science. In fact, a tendency to maxi-
mize disorder (entropy) in the universe is the driving force for the spontaneous act in chemistry. As for gases, they are indeed the chaotic state—molecules in constant motion, moving
randomly, with a range of velocities, colliding frequently.

Molecular motion is heat. On the average molecules move very quickly. In fact, the
average velocity of air molecules, around 1,000 miles per hour, is a little greater than the speed
of sound. That there is a connection between molecular speed and the speed of sound is no
surprise, for sound clearly derives from disturbances set up in the medium in which it propa-
gates. The average speed of molecules varies with temperature (the hotter it is, the faster the

molecules move) and with the mass of the molecules (light hydrogen cruises with the record speed of nearly 4,000 miles per hour at room temperature):

Deep in,
they're there, they're
at it all the time, it's jai
alai on the hot molecular fronton—
a bounce off wall onto the packed aleatory
dance floor where sideswipes are medium of exchange,
momentum trades sealed in swift carom sequences,
or just that quick kick in the rear, the haphaz-
ard locomotion of the warm, warm world.
But spring nights grow cold in Ithaca;
the containing walls, glass or metal,
are a jagged rough rut of tethered
masses, still vibrant, but now
retarding, in each collision,
the cooling molecules.
There, they're there,
still there,
in deep,
slow

.

One aspect of our experience seems to contradict the idea of molecules moving at sonic speeds. When a pleasant or foul-smelling odor is released in our environment, it does not reach us instantly. Some time passes before we smell it. The reason for this is that under typical atmospheric conditions, molecules travel but a very short distance before encountering an-other. A perfume molecule might set out at 500 miles per hour in our direction, but before it

*Greek Air* (1989)

has gone more than one ten-thousandth of a centimeter, wham, it's knocked sideways by an interfering nitrogen molecule from the air. Eventually molecules of perfume do find their way to our nose, but only by random, buffeted paths. That's diffusion. Now in outer space, where there's little in the way . . .

The title of the above poem is "Heat:Hot, as _____:Cold."

# ■ GREEK EARTH ■

"Dust you are, to dust you shall return." Or, in a more optimistic vein, all the richness of the immediate world—all creatures, plants, the synthetic bounty of our hands and minds, derives from the earth (and its atmosphere). The "system," which some have called Gaia, is almost closed as far as matter is concerned, but there is much influx of energy from the sun, and much diffuse heat emitted by the earth.

But what *is* the earth made of?

Well, it matters what pretty piece of real estate you choose. The charts below show the composition of the universe, the whole earth, and the earth's crust (the uppermost 20 km):[11]

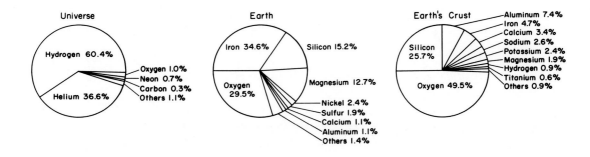

Universe

Hydrogen 60.4%
Helium 36.6%
Oxygen 1.0%
Neon 0.7%
Carbon 0.3%
Others 1.1%

Earth

Iron 34.6%
Silicon 15.2%
Magnesium 12.7%
Oxygen 29.5%
Nickel 2.4%
Sulfur 1.9%
Calcium 1.1%
Aluminum 1.1%
Others 1.4%

Earth's Crust

Aluminum 7.4%
Iron 4.7%
Calcium 3.4%
Sodium 2.6%
Potassium 2.4%
Magnesium 1.9%
Hydrogen 0.9%
Titanium 0.6%
Others 0.9%
Silicon 25.7%
Oxygen 49.5%

How different the earth is from the universe! Hydrogen, the nuclear fuel of stars such as the sun, and helium, the product of these great fires, dominate the stuff of the universe. At one time they too were present on the earth, but they have long ago escaped our atmosphere. Whatever is left of hydrogen is tied up mostly in one crucial molecule, water.

Note the vast amount of iron in the earth as a whole. It is largely in the planet's molten core, inaccessible. At the earth's surface, we have mostly minerals in rocks, including vast quantities of silicon oxides or silicates, and many aluminum-containing minerals as well. Much of the earth's oxygen is tied up in minerals, not in the atmosphere.

And in us? There is no chart here, but you know that water makes up 70% of the living human, and much of the remainder is an assortment of plain and fancy organic molecules, all containing carbon. Note how little carbon is in the crust of the earth—it's lumped in the 0.9% "other." This is a harbinger of the extraordinarily successful, from the point of view of life, chemistry of carbon.

All over the earth this wealth of elements seems to show itself to us in the form of amorphous-looking rocks, soils, plants, and flesh. But here and there miners came across something else: crystals. Think of shining a torch into a newly opened vug filled with quartz crystals, matte-gray galena, fluorite, purple shading to yellow, a few gem-like ruby sphalerite crystals! Who had made them that way?

The question is a natural one, because chaotic patterns are the norm in nature. Imagine aerial reconnaissance of a planet, this planet if you like, for intelligent life. The quickest guidepost to human-like activity is to be found in the constructed struggle with natural disorder. Local victories are found: plowed fields, buildings, a Vivaldi opera. So shiny, geometrical crystals might well be thought a sign of guided action. You may recall the premise explored in Arthur Clarke's *2001: A Space Odyssey.*

And they are organized, these near-perfect latticeworks of atoms marching in procession toward infinity. But they are also absolutely natural, a testimony to the symmetry of the underlying forces, unusual only in that their microscopic order also manifests itself in their macroscopic appearance. All those other seemingly inchoate things of this earth—granite, a fern—are also ordered within. Crystals just beckon more directly, prompting us to search for the order within.

*Greek Earth* (1989)

Mircea Eliade tells how among Australian aborigines shamans are initiated by being fed quartz crystals. These "stones of light" are pieces broken away from the heavenly throne. They allow the shaman to see, into space, into the soul, into matter.

*The Devil Teaches Thermodynamics*

My second law, your second law, ordains
that local order, structures in space
and time, be crafted in ever-so-losing
contention with proximal disorder in
this neat but getting messier universe.
And we, in the intricate machinery of our
healthy bodies and life-support systems,
in the written and televised word do declare
the majesty of the zoning ordinances
of this Law. But oh so smart, we think
that we are not things, like weeds,
or rust, or plain boulders, and so
invent a reason for an eternal subsidy
of our perfection, or at least perfectibility,
give it the names of God or the immortal
soul. And while we allow the dissipations

that cannot be hid, like death, and—in literary
stances—even the end of love, we make
the others just plain evil: anger, lust,
pride—the whole lot of pimples of the spirit.
Diseases need vectors, so the old call
goes out for me. But the kicker is that the struts
of God's stave church, those nice seven,
they're such a tense and compressed support
group that when they get through you're really
ready to let off some magma. Faith serves up
passing certitude to weak minds, recruits for
the cults, and too much of her is going to play
hell with that other grand invention
of yours, the social contract. Boring
Prudence hangs around with conservatives,
and Love, love you say! Love one, leave
out the others. Love them all, none will love
you. I tell you, friends, love is the greatest
entropy-increasing device invented by God.
Love is *my* law's sweet man. And for God
himself, well, his oneness seems too
much for natural man to love, so He comes up
with Northern Irelands and Lebanons . . .

The argument to be made is not
for your run-of-the mill degeneracy, my
stereotype. No, I want us to awake,
join the imperfect universe at peace with

*Greek Fire* (1989)

the disorder that orders. For the cold
death sets in slowly, and there is time,
so much time, for the stars' light to scatter
off the eddies of chance, into our minds,
there to build ever more perfect loves,
invisible cities, our own constellations.[12]

# ■ GREEK WATER ■

Water is by far the most common liquid in our environment (there is also alcohol, and vinegar, both usually encountered quite diluted with water, and mercury and . . . ). It's rare elsewhere in the universe, but the proportion of water in our bodies is close to that of water in a thin surface layer of the earth—70%.

Water is made up of $H_2O$ molecules, though for a good part of the nineteenth century people thought it was OH.

The $H_2O$ molecules, tiny, are bent. The HOH angle is 104.5°, the O-H distance 0.0000000097 cm. It's lucky that there are many of them, $6.02 \times 10^{23}$ (the number of atoms chemists have agreed to call a mole, give or take a few) in 18 grams, a small slurp. If the molecules of water were not bent, theorists predict that it would be a gas at room temperature. Life wouldn't be the same.

In gaseous water, steam, the molecules are bouncing around chaotically, moving near the speed of sound, not getting very far before colliding with each other. In liquid water, the molecules are also moving randomly, but they are much closer to each other, held together by fairly strong forces. In ice, solid water, these forces sculpt an ordered latticework. Strangely, ice is slightly less dense than liquid water. That's not what one would expect. This curiosity is responsible for the existence of aquatic life in wintry climes, for if ice were more dense than water, lakes would freeze from the bottom up.

*The hydrologic cycle.* The amount of water on earth is vast, yet most of it is unavailable to humans. Ninety-seven percent of the water is in salty seas, and four-fifths of the remainder is locked away in the ice caps. What fresh water is available to us depends on the evaporation of seawater and its precipitation on land.

It is estimated that of the 40,000 km³ of water per year that is within our reach through run-off from rain or snow over land, we in fact use about 3,000 km³, over 80% of that in agriculture, the remainder divided between domestic and industrial use.

*The uses of leaky vessels.* Were the earth truly impervious to water, our lives would be different. But the earth is leaky; groundwater that saturates soil and rock formations supplies much of our drinking water. Water moves down through sandy soil and fractured rock, along more solid rock layers, seeking discharge at lower levels in springs, wells, and wetlands. Along the way the water is cleaned, the biological waste removed by microorganisms, the solids separated mechanically. Agriculture, life depends on aquifers.

Water purification, especially the immense treatment systems of our cities, relies on filtration; the modern chemical laboratory also retains the characteristic shape of the alchemists' filter funnel. Controlled leakage, for that is what filtration is about, figures importantly in the transport of water and other molecules across biological membranes. In dialysis, a stream of blood circulates through an artificial kidney. The blood is separated from a solution by a semi-permeable membrane; waste products in the blood diffuse through the membrane into the solution and are discarded. The pores in the membrane are just large enough to allow the waste molecules to move through, but not the blood cells and other large biological molecules.

*Greek Water* (1989)

A general principle at work in societies *and* in the human cell is that economy of function is achieved by sequestering and communication. One segregates or specializes to accomplish a task; yet the partitions have to be leaky, just enough to allow real or molecular commerce. In Greek mythology, the Danaids' punishment for the unnatural act of killing their husbands was to fill, in perpetuity, leaky vessels. Perhaps this was symbolic of reintegration into the hydrological cycle—water in holey vessels.

# ■ THEORY AND PRACTICE ■

One Saturday morning I went into our library, as I'm wont to do on Saturdays. Right near the entrance there's a rack for the new journals that have come in that week. On Monday they're taken away, to another place in the library, to lie there for a year, until bound. The new journal rack is a place to look at what's new. The hundred-odd journals stare me in the face as I enter the library; they make me feel guilty that I might have missed something important. Once you get in the habit of reading them you cannot stop—you are caught in the web of the new. If you are really addicted, you come each day, as the journals are put out. When I was a beginner in chemistry, and had less pulling at me, that's what I did. Now I just channel my obsession, come in weekends to my pantheon of new molecules. The new journals are put out during the librarians' working week, ending Friday. They will be taken away Monday. This is the readers' interval.

On that day in 1989 I noticed in the *Journal of Organometallic Chemistry* an article by Margarita Rybinskaya from the Nesmeyanov Institute of Organoelement Chemistry in Moscow. She and her coworkers (mostly female; after a visit to her laboratory, Rita said that she hoped I noticed that our obligatory tea was served by the most beautiful chemists in the world) are experts at making certain molecules, called organometallics, of iron (Fe), ruthe-

62

nium (Ru), and osmium (Os), plus carbon and hydrogen. The newly synthesized molecule she reported on in her article is structure **1** below:

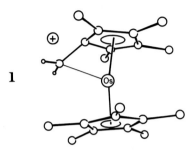

**1**

Note the Os atom sandwiched between two rings, each made up of five carbon atoms, each called a cyclopentadienyl. Note, too, the "extra" carbon at the top, distorted, leaning toward the osmium. That carbon is called a carbonium ion; the whole molecule bears a positive charge. The compound is colloquially called the osmocenyl carbonium ion.

That Saturday, as I looked at Rybinskaya's paper in the library, it occurred to me to think of the corresponding dication, structure **2**. In it both rings bear an external carbonium ion, two positive charges on the molecule as a whole, therefore the dication label. I don't draw the expected bending of the carbonium ions down (and up) toward the metal, just to make the overall geometry clearer.

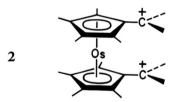

**2**

Now there's absolutely nothing original in that thought. If someone paints one leg on a chair red, one is led to thinking of painting two, or three, or all four that color. And the two legs can be opposite each other or adjacent. Much of science is like that. But there was a new problem in **2**, one of potential interest to a theoretician. The orientation, relative to each other, of the two carbonium ion centers may vary. In **2** the two "arms" are depicted on top

*Theory and Practice* (1990)

of each other, in a picture shown again in a side view and a schematic projection from the top in **3**. In **4** they are rotated 90° apart. The rotation is around the "vertical" axis, the one piercing the centers of the cyclopentadienyl rings and containing the metal atom. In **5** the arms are 180° apart.

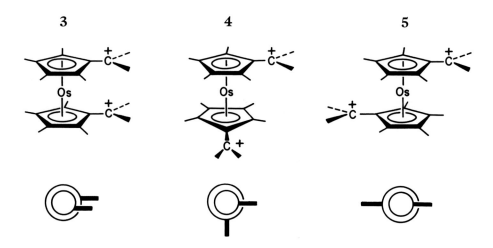

I sketched the disposition of the electrons on the osmium; I remember doing this in the library. The reasoning could be explained to a graduate student in inorganic chemistry in five minutes; if I do not reproduce it here, it is merely testimony to the fact that there is a reason why it takes five years to get a Ph.D. in chemistry. Anyway, in true "back of the envelope" fashion I came to the conclusion that the preferred geometry should have the external carbons rotated 45° or 135° relative to each other, as in **6** or **7**.

I came back to my office and wrote a letter to Rita. I would have written that letter even if I didn't know her—communication is really easy in science. But as it happens, I've known this outstanding Russian scientist for over a decade. I remember well a snowy December day when I came in on the last plane they would let land in Sheremetovo for three days. Rita and a driver met me, and we drove for hours through deep, drifting snow, in weather suitable for troikas, past militia outposts, no one else crazy enough to be on the road. It was the last day of a school in organometallic chemistry; I was delayed for good reasons. In the middle of the blizzard, Mother Russia and Grandfather Frost calling out to us with their wintry voices, we skid our way to a modern hostel, where . . . a home-made disco is under way, replete with a rotating, mirrored ball jury-rigged from lab supplies, American hard rock on tape, a hundred organometallic chemists dancing on their last night there, joining hands to sing a song by Bulat Okudzhava. While outside . . .

In my letter to Rybinskaya—in English because I'm afraid of writing Russian, even though I speak the language—I say something along the following lines: Fantastic that you've made the osmocenyl carbonium ion! Wouldn't it be interesting to make the dication? There are some predictions we could make about its structure.

Margarita replies, in a while (sad to say it takes six weeks for a letter to go between the United States and the former Soviet Union, either way—who is inefficient, who reads them, I don't want to know). She writes "Dear Roald" in English, but then she's afraid of how poor her English is (which she has to use in her papers, because, like-it-or-not, broken English has become the international language of science), so she continues in Russian: Great, we'll try; meanwhile you do your calculations, maybe we can publish our work together.

In my lab then was another Russian, Ruslan Minyaev from Rostov-on-Don University. I had met Ruslan in 1985 in Rostov. We had visited the old Don Cossack communities together, we swam in the strong currents of the Don. I liked the way Ruslan did his science, and how he cared for drawings of chemical structures, so I invited him to Cornell. Because he wasn't a party member, and because he was at a university, which is lower on the Russian totem pole than their Academy of Sciences, and because perestroika took its time to truly materialize, it was four years before he received permission to visit.

I asked Ruslan if he was interested in studying the dication which Margarita's lab was presumably hard at work making. I liked the idea of one Russian in Ithaca working theoretically on a molecule that another Russian in Moscow was synthesizing. And, it being a big country, the two Russians didn't previously know each other. I liked being a matchmaker.

In time the numbers spewed out of our computer, telling us that no, the molecule did not want to be in geometry **6** or **7**, as I had predicted. Insubordinate, it preferred instead **4**, the two carbonium ion centers rotated by 90° relative to each other. What right had this computer to tell me the molecule would not be doing what I so cleverly thought it should be doing? Well, Ruslan eventually came up with an explanation, one simple enough to make me want to kick myself for not having seen it earlier.

In the midst of this theoretical contretemps, Margarita writes: You'll be glad to learn that we've made the dication, just as you predicted (she knows the way to a theoretician's heart!). I turn the page to her drawing and there is **8**. Not the dication I had thought of, but another one, with sure enough two carbonium ions in the molecule, but both on one ring, instead of one each on two separate rings!

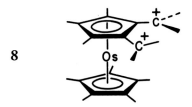

**8**

Rybinskaya had made what a chemist would call an isomer of the molecule I wanted, a molecule that was the same but not the same. In my earlier letter I had violated a basic chemical principle: I failed to supplement my words with a chemical structure, a little drawing. There was no fault in what Rita did. On the contrary, she had made a molecule no one had made before, a first! So while Ruslan scrambled to do theory on Rita's molecule, so that we could write a joint paper on it, I had to tell her: Please try to make the other one. She did try. It's harder; they're still trying.

# ■ THE CHEMIST ■

In describing what they do, scientists have by and large bought the metaphor of discovery and artists that of creation. The cliché "uncovering the secrets of nature" has set, like good cement, in our minds. But I think that the metaphor of discovery is effective in describing only part of the activity of scientists, and a smaller piece still of the work of chemists. The historical, psychological, philosophical, and sociological reasons for the ready acceptance of the metaphor deserve a closer look.

*History and psychology.* The rise of modern science in Europe coincided with the age of geographical exploration. Men set foot on distant shores, explored *terra incognita.* Even in our century, a man I was named after first sailed the Northwest passage and reached the South Pole. Voyages of discovery, maps filled in—those are powerful images indeed. So is penetration into a royal tomb full of glistening gold vessels. It's no surprise that these metaphors were and are accepted by (predominantly male) scientists as appropriate descriptors of their generally laboratory-bound activity. Is there some vicarious sharing of imagined adventures at work here?

*Philosophy.* The French rationalist tradition, and the systematization of astronomy and physics before the other sciences, have left science with a reductionist philosophy at its core. There is supposed to exist a logical hierarchy of the sciences, and understanding is to be defined solely in vertical terms as reduction to the more basic science. The more mathematical, the better. So biological phenomena are to be explained by chemistry, chemistry by physics, and so on. The logic of a reductionist philosophy fits the discovery metaphor—one digs deeper and discovers the truth.

But reductionism is only one face of understanding. We have been made not only to disassemble, disconnect, and analyze but also to build. There is no more stringent test of passive understanding than active creation. Perhaps "test" is not the word here, for building or creation differ inherently from reductionist analysis. I want to claim a greater role in science for the forward, constructive mode.

*Sociology.* Those philosophers of science who started out as practicing scientists have generally, I believe, come from physics and mathematics. The education of professional philosophers is likely to favor the same fields; quite understandably, there is a special role for logic in philosophy. No wonder that the prevailing ideology of reasoning in the underlying scientific areas of expertise of philosophers of science has been extended by them, unrealistically I believe, to all science.

What is strange is that chemists should accept the metaphor of discovery. Chemistry is the science of molecules (up to a hundred years ago one would have said "substances" or "compounds") and their transformations. Some of the molecules are indeed *there*, just waiting to be "known" by us, their static properties—what atoms are in them, how the atoms are connected, the shapes of molecules, their splendid colors—and in their dynamic characteristics—the molecules' internal motions, their reactivity. The molecules are those of the earth—for instance, simple water and complex malachite. Or of life—relatively simple cholesterol and more complicated hemoglobin. The discovery paradigm certainly applies to the study of these molecules.

But so many more molecules of chemistry are made by us, in the laboratory. We're awfully prolific. A registry of known, well-characterized compounds now numbers nearly ten million. These were not on earth before. It is true that their constitution follows underlying rules, and if chemist A had not made such-and-such a molecule on a certain day, then it is likely to have been synthesized a few days or decades later by chemist B. But it is a human

*The Chemist* (1989)

being, a chemist, who chooses the molecule to be made and a distinct way to make it. This work is not so different from that of the artist who, constrained by the physics of pigment and canvas, shaped by his or her training, nevertheless creates the new.

Even when one is clearly operating in the discovery mode in chemistry, elucidating the structure or dynamics of a known, naturally occurring molecule, one usually has to intervene with created molecules. I recently heard a beautiful lecture by Alan Battersby, an outstanding British organic chemist, on the biosynthesis of uroporphyrinogen-III. (Even in the trade, the name of this molecule is abbreviated as uro'gen-III.) It's not a glamorous molecule, but it should be: for from this precursor plants make chlorophyll, the basis of all photosynthetic activity. All cells use another uro'gen-III derivative in cytochromes for electron transport. And the crucial iron-containing, oxygen-carrier piece of hemoglobin derives from this small disk-shaped molecule.

Uro'gen-III, pictured below, is made from four rings, called pyrroles, themselves tied into a larger ring.

$A = CH_2COOH$
$P = CH_2CH_2COOH$

Note the markers A and P in each ring. They're in the same order as one goes around the ring (from about 10 o'clock), except for the last set, which are "reversed." So the markers read A, P, A, P, A, P, P, A.

How this natural molecule is assembled, within us, is clearly a discovery question. In fact, the four pyrrole rings are connected up, with the aid of an enzyme, into a chain, then cyclized. But the last ring is first put in "incorrectly," that is, with the same order of the A, P labels as in the other rings: A, P, A, P, A, P, A, P. Then, in a fantastic separate reaction sequence, just that last ring, with its attached labels, is flipped into position.

This incredible but true story was deduced by Battersby and his coworkers using a sequence of synthetic molecules, not natural ones, which were made slightly different from the natural ones. Each was designed to test some critical part of the natural process in the living system. Each was then treated under the physiological conditions to allow the sequence of the natural events to be traced out. Using molecules we've made, we've learned how nature builds a molecule that makes life possible.

The synthesis of molecules puts chemistry very close to the arts. We create the objects that we or others then study or appreciate. That's exactly what writers, composers, visual artists, all working within their areas, working perhaps closer to the soul, do. I believe that, in fact, this creative capacity is exceptionally strong in chemistry. Mathematicians also study the objects of their own creation, but those objects, not to take anything away from their uniqueness, are mental concepts rather than real structures. Some branches of engineering are actually close to chemistry in this matter of synthesis. Perhaps this is a factor in the kinship the chemist-narrator feels for the builder Faussone, who is the main character in Primo Levi's novel *The Monkey's Wrench*.

In the building of theories and hypotheses, even more than in synthesis, the act is a creative one. One has to imagine, to conjure up a model that fits often irregular observations. There are rules; the model should be consistent with previously received reliable knowledge. There are hints of what to do; one sees what was done in related problems. But what one seeks is an explanation that was not there before, a connection between two worlds. Often, actually, it's a metaphor that serves as the clue: "Two interacting systems, hmm . . . , let's model them with a resonating pair of harmonic oscillators, or . . . a barrier penetration problem." The world out there is moderately chaotic, frighteningly so, in the parts we do not understand. We want to see a pattern in it. We're clever, we "connoisseurs of chaos," so we find/create one. Had more philosophers of science been trained in chemistry, I'm sure we would have a very different paradigm of science before us.

Is art all creation? I don't think so. In substantial measure it is discovery, of the deep truths of what is also around us, often overlapping, but more often reaching outside the set of problems that science has set for itself to try to understand. Art aspires to discover, explore, unravel—whatever metaphor you please—the non-unique, chanced, irreducible world within us.[13]

# ■ SIMPLY BURNING ■

How did this wonderful hodgepodge of elements come to be? In the beginning there was a *big bang,* so it seems, and things were truly hot. A second later the temperature came down to a mere ten billion degrees C, and protons, neutrons, electrons could exist. In a few minutes a good bit of helium formed, by the reactions:

$$p + n = {}^2H + \gamma$$
$$^2H + {}^2H = {}^3H + p$$
$$^2H + {}^2H = {}^3He + n$$
$$^3He + n = {}^4He + \gamma$$
$$^3H + p = {}^4He + \gamma$$

The notation here is the following: p = proton, n = neutron, $\gamma$ = gamma radiation, very energetic light; in $^3H$, $^3He$, $^4He$, the total number of protons and neutrons precedes the name of the element as a superscript. $^2H$ is deuterium, $^3H$ is tritium, a radioactive form (isotope) of hydrogen.

> When God made the sun
> he lay back on his white
> sand beach, and reaching
> out, with both pale hands,
> into his space, he shaped
> there a sphere of hydrogen,
> God did, set it alight
> with his nuclear fire. He
> felt, God felt, its warmth
> on his soft hand. And
> it was good, it was his sun.

The temperature fell (not quite healthy for us yet) and so did the density of the original matter. It would have been nice to form the heavier elements right here, in the tail of the big bang, but apparently it couldn't be done until stars condense, and new reactions can run. The most important of these is the chain:

$$^1H + {}^1H = {}^2H + e^+ + \nu$$
$$^2H + {}^1H = {}^3He + \gamma$$
$$^3He + {}^3He = {}^4He + 2\,{}^1H$$

Here $e^+$ is a positron and $\nu$ a neutrino. This is hydrogen burning; the process provides the energy source for stars during the major part of their lives.

These nuclear reactions prevent further collapse of the star, stabilize it in size and luminosity. We still do not have the heavier elements, only hydrogen and helium. Now their synthesis begins:

$$^3He + {}^4He = {}^7Be + \gamma$$
$$^7Be + e^- = {}^7Li + \nu$$
$$^7Be + {}^1H = {}^8B + \gamma$$

*Simply Burning* (1991)

The neutrinos produced sail out of the star with ease. Deep in a mine there is an experiment trying to count these neutrinos, by using a tank of perchloroethylene, $C_2Cl_4$. The reaction

$$^{37}Cl + \nu = {}^{37}Ar + e^-$$

is how the neutrinos are detected, by measuring the radioactive argon formed.

> When God set about next
> to make the moon, he put
> his feet on the ice cap
> of Mars, and reached out
> again, seizing a piece
> of an old sun, and God
> threw it, like a snowball,
> at his earth. The earth
> rocked, and so the moon,
> God's moon, came to be. He
> felt its reflecting light,
> and it was good, his moon.

On to the heavier elements. A so-called CNO cycle makes some isotopes of C, N, and O. But the main source of the heavier elements is helium burning, which steps in when a star depletes its hydrogen and comes to be mainly composed of He. The star contracts again, the temperature rises, and reactions such as

$$3\,^4He = {}^{12}C + \gamma$$

set in. When the helium is exhausted, the carbon burns. This can happen only in stars heavier than our sun. $^{16}O$ also begins nuclear reactions here, eventually forming Si, P, S, Na, and Mg. In the final stage of energy production in a star, silicon burns, producing the elements around iron in the periodic table. Still heavier elements come from successive capture of ambient neutrons and protons.

If our home star isn't big enough to have formed Si or Fe, where do these elements, so abundant on earth, come from? From the spectacular disruption of a heavy star in a nuclear explosion, a supernova. Unimaginable (no, terrible, strange, but quite imaginable) events happen, quickly. The Crab Nebula is the remnant of a supernova recorded by Chinese astronomers in 1054. All the elements, heavy and light, in that mature star are blown away. The observed abundances of the elements in our solar system may be accounted for by assuming one-ninth of them were generated in a supernova. The rest is the outcome of normal labor by our sun.

When the time came for God
to people this blue earth,
he stood knee-deep in paddy
and sea, and, dear God, he
didn't make people in his
image, but just reached out
his now sunburnt hands
to plant a mitochondrion,
here a squid's eye, a seed
of rice. Hazard he gave them,
rules, God's time, and soon
enough, the creatures came,
spoke. It was good, the word
between God and his people.

# ■ THE PHILOSOPHER'S STONE ■

Alchemy presents a real problem for modern chemistry. Here is one prominent stream from the past into the present, a protochemistry as much as are metallurgy, cosmetics, pharmacology, fermentation, ceramics, and dyeing. At the same time alchemy had a philosophical framework attached to it, one of maturation and transformation of chemical substances, and of the alchemist or *artifex* himself. Because the goal of change was the noblest metal, gold, or the perfect state of eternal life and health, and because that goal was in fact impossible to achieve, charlatanry crept in. Philosopher's stones and elixirs of life were faked.

What modern chemists would like to do is to keep the protochemistry, throw away the specious (to the chemists) conceptual base, and retain just enough of the swindle, nicely distanced, to evoke humor. In other words, to build a mythology, which humans or societies are wont to do. But this wishful sanitization of the past won't work. The parts of alchemy are parts of a whole. There would have been no alchemy (and, I believe, the coming of modern chemistry would have been delayed by a century) had it not been for those suspicious philosophical underpinnings of the alchemist's work.

Alchemies developed independently in China, India, Egypt and Greece. In its classical European form the alchemist's work was a series of transformations of matter, described, for instance, in Latin by Josephus Quercetanus in 1576 as:[14]

*The Philosopher's Stone* (1989)

1. *Calcinatio*

2. *Solutio*

3. *Elementorum separatio*

4. *Coniunctio*

5. *Putrefactio*

6. *Coagulatio*

7. *Cibatio*

8. *Sublimatio*

9. *Fermentatio*

10. *Exaltatio*

11. *Augmentatio*

12. *Proiectio*

The mix of chemistry and thought is clear in the very terms themselves and, as Carl Jung has perceptively pointed out, so is a psychological dimension. Matter was being transformed, and the alchemist was transformed as well. Jung traces the striking similarity between the alchemist's language and symbols and that of archetypical dreams, the work of the psyche.

Here is a typical alchemical passage, probably of Arabic origin:

> But when we marry the crowned king with the red daughter, she will conceive a son in the gentle fire, and shall nourish him through our fire. . . . Then he is transformed, and his tincture remains red as flesh. Our son of royal birth takes his tincture from the fire, and death, darkness, and the waters flee away. The dragon shuns the light of the sun, and our dead son shall live. The king shall come forth from the fire and rejoice in the marriage. The hidden things shall be disclosed, and the virgin's milk be whitened. The son is become a warrior fire and surpassed the tincture, for he himself is the treasure.

You can see the problem. Underneath the symbolic language a set of chemical transformations takes place. Color changes, then as now, are a fascinating part of the experimental enterprise. In the context of modern chemistry, it's interesting that the emphasis is not on the static structure of matter but on its dynamics. You could spend a goodly amount of time reconstructing from the vague language—the precision of Lavoisier and Berzelius is at least half a millennium away—what reagents "the crowned king," "the red daughter," and "the son of royal birth" are. But it's impossible to miss the philosophy of eternal change achievable by human intervention that provided the incentive for lifetimes of experimentation. One way to look at it is that it was much like modern chemical theory: a rationale consistent with its times, not necessarily right, but rich, a goad to interpretation and action. Jung sees in this tale the Christian drama of Easter.

It is also clear that the discipline was deeply esoteric. Secrets were to be won, in apprenticeship and hard work, then to be concealed. A striking image is that of the alchemist and his *soror mystica* (a female coworker, real and spiritual, interesting in the context of the times), at the end of their work, signing to the reader to keep the knowledge secret. The modern chemical literature, the context of open publication and reproducibility (always contending with proprietary secrecy) had not yet been invented. The alchemical work, especially given its ambitious ultimate aim, lent itself to hocus-pocus.

Mircea Eliade, a historian of religion, has written a remarkable book, *The Forge and the Crucible,* which traces the relationship between religion, metallurgy, and alchemy. In his beautiful concluding chapter, Eliade makes the haunting observation that the goal of the alchemist was to hasten the "natural" evolution of metals from base to noble, and to secure a similar transformation of the body, from sick to healthy, from mortal to eternal. The alchemists failed, in the end, and were replaced by modern chemists and physicians, who, denying a connection all the way, have achieved, through catalysts, composites, and pharmaceuticals, a very large part of the alchemist's original goal.

# ■ P H L O G I S T O N ■

Chemistry gone astray, deluded for a hundred years by a false theory, that of phlogiston. Such a convenient set-up for the rationalizing revolution to come at the end of the eighteenth century!

The reality was that of an incorrect but fruitful idea that served well the emerging science of chemistry. At the heart of the theory was fire. This was consistent with the alchemical tradition. And fire differed from the other elements, much as a verb differs from a noun.

Chemistry is the science of molecules (earlier one would say substances) and their transformations. Some of the changes are spontaneous, proceeding under ambient conditions. Some must be driven with an input of energy. That energy may come in the form of heat or light or electricity. But it took another two centuries to recognize these; in the 1600s heat, and its source, fire, seemed the only obvious generative principle—what was needed to transform wheat into bread, iron ore into steel.

The idea, growing out of the work of Johann Joachim Becher and Georg Ernst Stahl, was that the essence of fire was a substance called phlogiston. When matter burned, it gave off phlogiston. Wood was full of it, ashes empty. Hematite was iron lacking phlogiston. With much fire, much inflow of phlogiston (from coal, rich in the principle), it could be converted to useful iron, which in turn gave off its phlogiston when it rusted. Oxygen itself, when dis-

*Phlogiston* (1990)

covered by Priestley (and independently, before him, by Scheele), was termed "dephlogisti-cated air" because it supported combustion.

Phlogiston theory worked very well, substituting a "not A," or "minus A," wherever there was an A. For burning or rusting, the critical A is oxygen, as Lavoisier realized (though he missed that it could be sulfur or some other element). So rust is not iron minus phlogiston, but a compound of iron and oxygen. And burning is a combination with oxygen, heat and light given off, instead of a loss of phlogiston. As long as one was interested in the overall process, an argument could be shaped on either presence or absence. For instance, the inge-nious connection between burning and rusting, not obvious, was made before the oxygen theory appeared.

The standard argument against phlogiston, only made more precise by Lavoisier, was that substances gained weight upon rusting and upon burning (if it was organic matter that burned, you had to realize that a gas, carbon dioxide, was given off, and include that in the mass balance). So how could something, phlogiston, be given off if the weight increased? This did not bother the proponents of the theory as much as we think it should, because they had a holistic picture of chemistry as a science concerned with overall or intrinsic qualities. Weight did not seem to them worth worrying about, not yet. In 1781 Richard Watson wrote: "You do not surely expect that chemistry should be able to present you with a handful of phlogis-ton, separated from an inflammable body; you may just as reasonably demand a handful of magnetism, gravity or electricity to be extracted from a magnetic, weighty or electric body; there are powers in nature, which cannot otherwise become the objects of sense, than by the effects they produce, and of this kind is phlogiston."[15]

In the end they were wrong. Their theory was replaced by a new language, a new em-phasis, in fact concentrating on mass relationships, that worked better. Adherents to the phlo-giston theory for a while complained that the new methodology failed to *explain,* that it substituted measurement without comprehension for a framework of understanding. In a way they were right. But in time the new chemistry took hold, and found, first in the atomic theory and molecular weights, then in the new quantum mechanics, the correct theoretical founda-tion. But the explainers with fire, heirs to Prometheus, deserve more credit than history af-fords them.

*Ponder Fire*

I wonder if phlogiston theorists
were lovers, if it began when they
were set off, like the brown grass

on the hills a little north of here.
It takes so little, a touch, to burn.
They had it right, sly Becher

and Stahl, the principle is fire.
Wood, coal, and lovers, and metal
too are rich in it, it's what's

expelled in a flame. And the stuff
left behind, spent ashes (and they
were right too in the slow burn

of rust) is emptied, lax, the head
of a long untuned drum. An inconstant
agent at the heart of this plausible

theory, sometimes free, sometimes
much combined with the base, antsy
to move out, but often held, dearly.

Its loosing can banish weight, as you
coming on me do. It can add stones,
the thought this consuming day will end.[16]

# ■ AIR OF REVOLUTION ■

Two centuries ago the French Revolution shaped a historical exclamation point. It changed the world's perception—of absolute monarchy, of nationhood, of the rights of man. It also coincided, that best of times, that worst of times, with a revolution in chemistry. The French Revolution also affected critically, even mortally, the lives of at least two of the three men who crafted the change in chemistry.

These three were Joseph Priestley, Carl Wilhelm Scheele, and Antoine Lavoisier. The chemical revolution was as complicated and multiform as the French one; to assign it a precise date is as simplistic as to think of the world changing on the day a nearly empty Bastille fell. The components of that chemical revolution were the discovery of oxygen and the true nature of burning, the attendant end of the phlogiston theory, and the essential introduction of precise quantitative measurement in chemistry.

*The intellectual setting.* A dominant chemical theory of a century's standing, Stahl and Becher's phlogiston. Phlogiston was posited as the principle of fire itself, given *off* in the process of combustion of any substance. It described well, metaphorically, the spent nature of burnt matter.

*Air and burning.* The indispensability of air for combustion was well known. There were strong hints that air was a *mixture* of gases, as hard to believe as that was of that seem-

*Air of Revolution* (1990)

ingly most homogeneous substance. A candle flame or a live animal exhausted the life-principle of the air after consuming only a fifth of it.

*Measurement.* Critical by then, indeed a hallmark of astronomy and physics, only slowly perceived as essential to chemical experimentation.

Joseph Priestley was born near Leeds, England, in 1733. He was a religious dissenter. To believe, as he, one of the founders of the Unitarian Church, did, that " . . . Jesus was in nature truly and solely a man, however highly exalted by God," was not likely to endear him to prevailing Anglican orthodoxy. Nor did his siding with the claims of the American colonists in a contemporary independence struggle, nor his quiet but determined marking of the achievements of the French Revolution, his admiration for the rights of man. Priestley's meeting house was burned by a mob on the second anniversary of Bastille Day in 1791; he fled to America three years later, there to spend the last ten years of his life.

A latecomer to chemistry, Priestley carried out his scientific experiments at home or at a public brewery nearby. He devised a neat way to heat intensely substances confined to a glass container, by using the lens-concentrated heat of the sun. A red powder, *mercurius calcinatus per se* (made by heating mercury in air, now known as mercuric oxide, $HgO$), heated by a burning lens on August 1, 1774, gave off a gas that supported burning and respiration in mice and proved to be the vital component of air. Priestley called it dephlogisticated air (for it encouraged fire, desired phlogiston), but what he had done was to make oxygen.

Priestley published his discovery quickly. This statement describes his attitude toward experimentation and publication: ". . . when I made a discovery, I did not wait to perfect it by more elaborate research, but at once threw it out to the world, that I might establish my claim before I was anticipated. I subjected whatever came to hand to the action of fire or various chemical reagents, and the result was often fortunate in presenting some new discovery."

Actually the gas had been discovered several years earlier by Carl Wilhelm Scheele. Scheele was born in 1742 in Stralsund, one of the great Hanseatic cities in Pomerania, on the German north coast. The city was still in Swedish hands in that year, nearly a century after it had become Swedish after the Thirty Years War. Scheele, probably German by origin, was apprenticed to an apothecary in Gothenburg at fourteen, and he remained in that calling in Sweden all his life. In much of Europe a pharmacist is called a chemist, testimony to the close historical ties of the two professions. Scheele was a truly wonderful chemist, the discoverer of many acids—of hydrogen cyanide, hydrogen fluoride, and hydrogen sulfide (all of which he sniffed and tasted, no doubt contributing to his untimely death)—and of the elements manganese, oxygen, molybdenum, and chlorine. However, credit for discovering the elements went elsewhere, an instructive story in each case.

Scheele made oxygen by decomposing a variety of salts, including the same mercuric oxide that served Priestley. He did it, bootlegging time as an apothecary's assistant, one to four years before the English minister did. Scheele called oxygen *eldsluft,* fire air, or *aër nudus,* or *aër purus.* His discovery was well known to the Uppsala community, but in putting it in the written or printed record, Scheele ran into trouble. First he took his time (unlike Priestley) to write a book. Then he faced a procrastinating, unreliable publisher—nothing new about that. And finally he waited for a tardy preface by an authority, the great Swedish chemist of his time, Torbern Bergman. The book did not appear until summer 1777, two years after Priestley's publication, at least four years after Scheele's discovery. Incidentally, Scheele, like Priestley, was a staunch advocate of the phlogiston theory.

Antoine Laurent Lavoisier was born a year after Scheele. Hardworking and versatile, he brought precise measurement to chemistry. The balance of science, like the balance in the hands of Justice, was to be his instrument—and not just one balance. Lavoisier had several, each carefully suited to cover a range of weights, each increasing in precision and accuracy. Through a series of careful weighings and experiments that ingeniously isolated the system under study (a diamond in a bell jar, his laboratory assistant in a silk bag, mercury heated to boiling), he showed the indestructibility of matter. Things certainly change, but nothing is lost, nothing created in a chemical reaction.

Priestley came to Paris in October 1774 and told Lavoisier of his experiments making dephlogisticated air. Lavoisier repeated them, made them more precise. He was the first to realize clearly that oxygen was the essential agent in burning; he was also the one who named it "oxygen" (from Greek, meaning "acid former"). Combustion (and respiration) was combination with oxygen, with attendant weight gain. The phlogiston theory, trying to wriggle out of this weight gain by postulating negative mass for the fire-principle, was just untenable—air would not support it. Lavoisier gave Priestley precious little credit. I suspect that Priestley's discovery (interpreted by Priestley in terms of the faulty phlogiston theory—the political radical was a chemical conservative) was a small but essential increment to a framework Lavoisier had already been building for some time. He could not quite bring himself, not yet, to articulate his own theory. And then he had difficulty in dealing with Priestley's less systematic discovery that pushed his, Lavoisier's, knowledge past the point of understanding, to daring to express that understanding.

It's interesting to note that Scheele communicated his discovery of oxygen in a personal letter to Lavoisier in early October 1774. It was a competitive science, even then.

Ambitious Lavoisier married Marie Anne Pierrette Paulze de Chastenolles when she was thirteen; she later studied art with J. L. David. There is a striking double portrait by David of M. and Mme. Lavoisier, now in the Metropolitan Museum of Art. Mme. Lavoisier is the dominant figure in the painting. Indeed, she is a remarkable figure in her own right. She illustrated Lavoisier's books and helped him run his laboratory. Eleven years after his death she married another charismatic figure, the American/British/Bavarian adventurer and scientist, Benjamin Thomson, Count Rumford. Through his marriage, Lavoisier came to buy a share in the Ferme Générale, the *ancien régime* tax collection agency. For that he was guillotined at the age of fifty-one in the terror of 1794. Perhaps the fact that he had earlier publicly exposed as faulty Jean Paul Marat's claims as a scientist also played a part in this tragedy; Marat's indictment of Lavoisier is vicious.

What is the lesson of these three biographies linked by the life-giver oxygen? That science is international? That it depends on human ingenuity that can operate in apothecary shops and breweries? That open communication of new findings is critical? That claims of priority tell us something about human nature, and the essential reproducibility of the underlying facts? That one can do marvelous science within a wrong theoretical framework? All that and, most important, that chemistry is firmly embedded in the context of society. It may wish to isolate itself in glass-glittery laboratories. But the world butts in, at the beginning, in the middle, at the end of life.

Where Scheele worked was determined by a struggle for European power one hundred years before his birth. How he lived (and that he could do chemistry) derived from the rules of the ancient profession of apothecaries. He moved from Gothenburg to Malmö to Stockholm to Uppsala to the hardly cosmopolitan Köping, all in search of a living. He lost (only for a time) his valid claim to have discovered oxygen because of the workings of patronage and the publishing trade.

Priestley spoke for social change, for placing political power in the hands of the people, and saw his scientific work disrupted as a consequence. He invoked his social conscience in the context of his chemical theories, and wrote to Claude Berthollet: "As a friend of the weak, I have endeavoured to give the doctrine of phlogiston a little assistance."

And Lavoisier, who discovered "no new body, no new property, no natural phenomenon previously unknown" (to quote Justus von Liebig), yet had the greatest influence of this trio on our science, he, Lavoisier, died at the hands of the perversion of the revolution that Priestley supported, and for which the latter was hounded out of his native country.

Nothing is gained, nothing created. Yet, also, nothing is simple. Neither the burning of a candle, nor the breathing of a mouse. Even less so the question of who really discovered oxygen, why the French Revolution changed, what if Lavoisier had lived, and Marat had been a better scientist.

# ■ INTERMEDIARY ■

*Fritz Haber*

invented a catalyst to mine cubic miles
of nitrogen from air. He fixed the gas
with iron chips; German factories coming
on stream, pouring out tons of ammonia,

fertilizers, months before the sea-lines
to Chilean saltpeter and guano were cut,
just in time to stock powder, explosives
for the Great War. Haber knew how catalysts

work, that a catalyst is not innocent, but
joins in, to carve off the top or undermine
some critical hill, or, reaching molecular
arms for the partners in the most difficult

stage of reaction, brings them near, eases
the desired making and breaking of bonds.

*Intermediary* (1989)

The catalyst, reborn, rises to its match-
making again; a cheap pound of Haber's

primped iron could make a million pounds
of ammonia. Geheimrat Haber of the Kaiser
Wilhelm Institute thought himself a catalyst
for ending the War; his chemical weapons

would bring victory in the trenches; burns
and lung cankers were better than a dum-dum
bullet, shrapnel. When his men unscrewed
the chlorine tank caps, and green gas spilled

over the dawn field at Ypres, he carefully
took notes, forgot his wife's sad letters.
After the War, Fritz Haber dreamed in Berlin
of mercury and sulfur, the alchemists' work

hastening the world, changing themselves.
He wondered how he could extract the millions
of atoms of gold in every liter of water,
transmuting the sea to the stacked bullion

of the German war debt. And the world, well,
it *was* changing; in Munich one could hear
the boots of brown-shirted troopers, one paid
a billiard marks for lunch. A catalyst again,

that's what he would find, and found—himself,
in Basel, the foreign town on the banks of his
Rhine, there he found himself, the Protestant
Geheimrat Haber, now the Jew Haber, in the city
of wily Paracelsus, a changed and dying man.

# ■ A HANDS-ON APPROACH ■

Of all the tensions that move the world, the most basic is that of identity and difference—the same and not the same. So it is in chemistry, as the story of chirality, or handedness, reveals. Some molecules exist in distinct mirror-image forms, subtly related to each other as a left hand is to a right. Many, but not all, of the properties of such mirror-image molecules are the same—they have identical melting points, colors, and so on. But some properties differ, often critically so. This is, for instance, true of their interaction with other handed molecules, such as the ones we have in our bodies. So enantiomers (the name given the distinct handed forms of a chiral molecule) may have drastically different biological properties. One may taste sweet, the other tasteless. The mirror-image form of morphine is a much less potent pain reliever.

Our knowledge of chirality begins in 1848 with a twenty-six-year-old Louis Pasteur, before he studied microorganisms, or invented pasteurization, or a vaccine for rabies. He became interested in optical rotation and linked it to a curious problem of the non-identity of two compounds that should have been identical.

It is in the nature of nature that its fuzzy details and seeming obscurities are clues to the world within. Optical rotation seems a curiosity even today. It deals with the ability of some substances to rotate the plane of polarized light, and this was discovered in France in the early nineteenth century. Normal light is a wave, electric and magnetic fields propagating in space.

It's possible to filter out polarized light, light in which the fields are now restricted to oscillate in a plane. You've seen polarizers—in sunglasses and in airplane windows—they are such filters.

A puzzle arose in the chemistry of another important part of French culture, wine making. One often sees colorless crystals, perhaps growing on the cork, in some white wines. These are a product of wine making (and they are much more abundant on the inside of wine casks and fermentation vessels!), a salt of tartaric acid. The naturally occurring material is in fact optically active, as are most biological molecules. In another stage of the fermentation process, a substance called racemic acid was isolated. It was made up of precisely the same atoms as tartaric acid but did not rotate the plane of polarized light. It was optically inactive—the same, but not the same . . .

Pasteur recrystallized a salt of racemic acid. Looking at the crystals under the microscope, he noted that they came in two varieties. Painstakingly, with tweezers, he separated the mirror-image crystal shapes, left-handed to one side, right-handed to the other. When dissolved separately, the solutions of the two crystal forms rotated the plane of polarized light in opposite directions. And one was identical to the naturally occurring tartaric acid.

Racemic acid is a 1:1 mixture of optically active tartaric acid and its enantiomer. These substances were not only differentiated in their crystal form; they were also optically active in solution. This must mean that the handedness is there not only in the big crystals but resides deeper, in the tiny molecules floating in solution.

One of the dramatic moments in the history of chemistry occurs when the dean of French optical rotation studies, J. B. Biot, skeptical of Pasteur's report, summons Pasteur to repeat his experiment in his laboratory. Biot prepares the salt of racemic acid according to Pasteur's prescription, Pasteur separates the crystals under the microscope, right there, under Biot's eyes. Biot dissolves the small samples of segregated crystals, measures their optical rotation himself. The essence of reliable knowledge, a reproducible experiment!

It took a quarter of a century, and the work of two other young chemists in their twenties, J. H. van't Hoff in Leiden and J. E. Le Bel in Strasbourg, to explain, in molecular detail,

*A Hands-on Approach* (1990)

what is behind optical activity. They proposed that carbon atoms are "tetrahedral," meaning that the four bonds that carbon forms are along the directions of a regular tetrahedron:

Our notation here is that standard (primitive) visual code chemists use for describing three-dimensional structures: a solid line is in the plane of the paper, a dashed line "in back" of that plane, a wedge in front.

Now consider the possible existence and identity of mirror-image forms, given the tetrahedral geometry of carbon. If you have one, or two, or three different substituents around a carbon atom (and that's what chemistry is about, changing one piece of a molecule for another), then the mirror image is identical to the molecule mirrored. Not so for *four* different groups on carbon, as shown below:

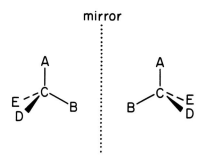

The molecule at left is *not* identical to the one at right. The only way to convince yourself of that fact is to try to superimpose the mirror images. If you put A and B on top of each other, E and D will be out of place. If you superimpose E and D, A and B won't fit.

What does all this have to do with hands? That may not be at first obvious, but the essential descriptors of a hand are thumb, pinkie, palm, back. They play exactly the same role as A, B, D, and E, the chemical groups that differentiate enantiomers. There is much more detail to a hand (fingerprints, your life?) and so there is to a molecule. But the topological essence of a hand or a molecule with one carbon is described by four markers not in a plane.

How does one separate mirror-image molecules from each other? The sorting of crystals, which is what Pasteur did, doesn't work very often. Another method, also devised by Pasteur, is to feed a living organism the mixture of enantiomers. The bacteria typically metabolize a molecule of one handedness, excrete the other. But there are many molecules even a bacterium won't eat. Here is a scenario from an unmade Antonioni film illustrating the most common method of what is called "optical resolution":

> You are about to enter a pitch-black room filled with mannequin parts, left and right hands. If you are not able to separate them, something terrible will happen to you. No problem. You begin to shake hands with the myriad paste mannequin hands. You put to one side the ones you can't shake hands with comfortably, to the other side the misfit left hands.

In resolution, a handed reagent is added to the mixture of left- and right-handed molecules. It forms two physically distinct compounds: the composite of a right hand shaking a left mannequin hand is different in shape from, and not the mirror image of, a right hand shaking a right hand. These composites are separable, they have different properties. They are separated, and then, in another reaction, knocked apart to yield the components.

Thalidomide, a sedative of the 1960s with terrible side effects of causing fetal malformation, was marketed as a mixture of its mirror-image form. But only one enantiomer causes malformations, not the other one.

# ■ CHINESE ELEMENTS ■

The first ingredient tossed in the witches' brew in the fourth act of *Macbeth* is hailed as follows:

> Toad, that under the cold stone,
> Days and nights hast thirty-one
> Sweltr'd venom sleeping got,
> Boil thou first i' the charmed pot!

Toad venoms indeed have been known for thousands of years. Remarkably they are also indispensable folk drugs in China and Japan, while nearly identical molecules are used by fireflies as protective chemicals. Their active principles are promising today not only for their stimulant action on the heart but also as anti-viral agents.

Despite a long tradition of scholarly investigation and a commercial, practical society, the protochemistry within Chinese alchemy never shed its philosophical overlay. If transformation is the heart of alchemy—that of base metals to gold, the alchemist into a better person, a sick body into a healthy one—then that mode of change that healed was dominant in Chinese alchemy. In a quest for the elixir of life, many emperors died prematurely, ingesting poisonous brews of orpiment and cinnabar. But native medicine flourished, and the Chinese

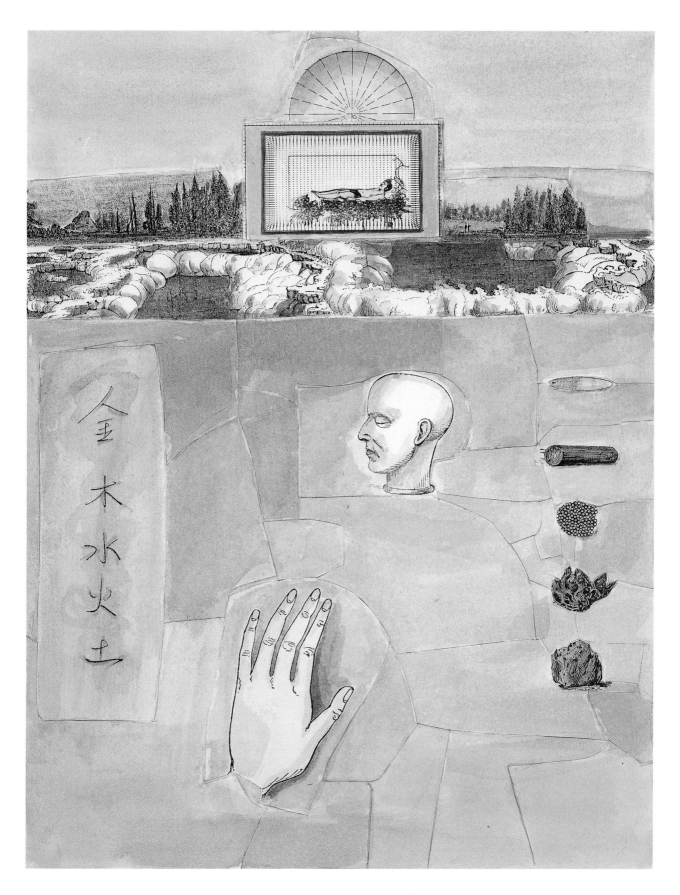

*Chinese Elements* (1989)

*materia medica* grew to fifteen hundred drugs of vegetable origin, five hundred or so from animal sources.

In the way that toxicity and benefit always seem to mix (so teaching us the inherent fallacy of branding molecular villains), the dried venom of the toad *Bufo bufo gargarizans* (Ch'an Su) was long used for the treatment of heart disease in China. In this century the many components in the extract of *Bufo* toad venom were isolated and their activity studied. One of the primary actors is a representative of the bufadienolide class, a molecule called bufalin.

Bufalin is a steroid, part of a rich class of pharmaceuticals, related to many hormones in the body. In the toad the molecule is probably assembled on a chemical sidetrack of a common, efficiently executed railroad yard of biochemical necessity. The same chemistry, or something near to it, is in all life.

It was then just a matter of time until it was found elsewhere. Tom Eisner, Jerry Meinwald, and some friends learned that hungry thrushes, broadly insectivorous, rejected certain types of fireflies. On close examination the distaste (to marauding birds) was traced to a class of molecules they called lucibufagins. Except for a little molecular tuning, an oxygen here or there, these are very close to the toad venom bufalin and its relatives, the bufadienolides.

And now the lucibufagins have been patented for their anti-viral activity! Questions, questions, new layers of questions: Even if we understand convergent molecular evolution, we remain ignorant of just exactly how bufalin deters the firefly's and toad's predators, how it affects an aching heart, or a virus, how it's decided that just this steroid may break through the chemical surface in toad and firefly, but be lacking in all the richness of the witches' brew of life in between.

# ■ BLOOD COUNTS ■

Look at the molecule below, structure **1**.[17] It seems there's nothing beautiful in its involuted curves, no apparent order in its tight complexity. It looks like a clump of pasta congealed from primordial soup or a molecular model builder's nightmare. The molecule's shape and function are enigmatic (until we know what it is!). It is not beautifully simple.

1
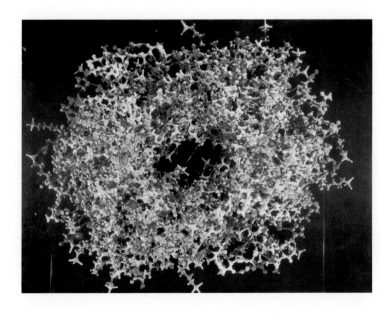

Complexity poses problems in any aesthetic, that of the visual arts and music as well as chemistry. There are times when the Zeitgeist seems to crave fussy detail—Victorian times, the rococo. Such periods alternate with ones in which the plain is valued. Deep down, the complex and the simple coexist and reinforce each other. Thus the classic purity of a Greek temple was set off by sculptural friezes, the pediments, and the statues inside. The clean lines and functional simplicity of Bauhaus or Scandinavian furniture owe much to the clever complexity of the materials and the way they are joined. Elliott Carter's musical compositions may seem difficult to listen to, but their separate instrumental parts follow a clear line.

In science, simplicity and complexity always coexist. The world of real phenomena is intricate, the underlying principles simpler, if not as simple as our naive minds imagine them to be. But perhaps chemistry, the central science, *is* different, for in it complexity is central. I call it simply richness, the realm of the possible.

Chemistry is the science of molecules and their transformations. It is the science not so much of the hundred elements but of the infinite variety of molecules that may be built from them. You want it simple—a molecule shaped like a tetrahedron or the cubic lattice of rock salt? We've got it for you. You want it complex—intricate enough to run efficiently a body with its ten thousand concurrent chemical reactions? We've got that too. Do you want it done differently—a male hormone here, a female hormone there; the blue of cornflowers or the red of a poppy? No problem, a mere change of a $CH_3$ group or a proton, respectively, will tune it. A few million generations of evolutionary tinkering, a few months in a glass-glittery lab, and it's done! Chemists (and nature) make molecules in all their splendiferous functional complexity.

Beautiful molecule **1** is hemoglobin, the oxygen transport protein. Like many proteins, it is assembled from several fitted chunks, or subunits. The subunits come in two pairs, called α and β. Incredibly, these actually change chemically twice in the course of fetal development, so as to optimize oxygen uptake. The way the four subunits of hemoglobin mesh, their interface, is requisite for the protein's task, which is to take oxygen from the lungs to the cells.

One of the hemoglobin subunits is shown in structure **2**.[18] It's a curled-up polypeptide chain carrying a "heme" molecule nestled within the curves of the chain:

*Blood Counts* (1990)

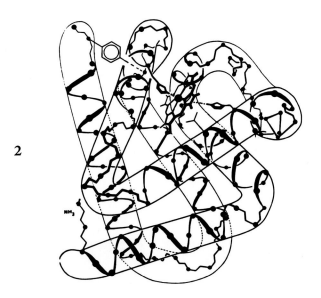

2

All proteins, not just hemoglobin, contain such polypeptide chains (see structure **3** for a schematic formula), which are assembled in turn by condensation of the building block amino acids, shown in structure **4**.

3

4

These come in about twenty varieties, distinguished by their "side chains" (R in structures **3** and **4**). A typical protein, the hemoglobin β-chain is made up of 146 amino acid links. Here is *richesse,* reaching out to us! Think how many 146-link chain molecules there could be, given the freedom to choose the side chains in twenty possible ways. The incredible range of chemical structure and function that we see in those tiny molecular factories, enzymes, or in other proteins, derives from that variety. The side chains are not adornment; they make for function.

The protein folds, the diversity of the side chains, provides opportunity; the particular amino acid sequence enforces a specific geometry and function. Extended pieces of hemoglobin curl in helical sections, clearly visible in structure **2**. At other places the chain kinks, not at random, but preferentially at one amino acid, called proline. The globular tumble of helical sections, nothing simple but functionally significant, emerges.

Significant in what way? To *hold* the molecular piece that binds the oxygen, and to *change*, in a certain way, once the oxygen is bound. Whenever I think of the process I see before my eyes the way the puck seemed to fit the glove of Ken Dryden, at Lynah Rink at Cornell, before this great hockey goalie joined the Montreal Canadiens. The $O_2$ winds its way into a just-right pocket in the protein, and binds to the flat, disk-shaped heme molecule. Heme's structure is shown in number **5**. The oxygen binds, end-on, to the iron at the center of the heme.

**5**

As it does, the iron changes its position a little, the heme flexes, the surrounding protein moves. In a cascade of well-engineered molecular motions, the oxygenation of one subunit is communicated to another, rendering that one more susceptible to taking up still another $O_2$.

That bizarre sculpted folding has a purpose, in the structure and function of a molecule critical to life. All of a sudden we see it in its dazzling beauty. So much so that it cries out: "I've been designed. For this task, I'm the best that can be." Or, if you're so inclined, it testifies to a Designer.

Beautiful? Certainly. The best, fashioned to a plan? Hardly. It only takes a moment to get us back to earth, a few bubbles of CO, the lethal, odorless product of incomplete combustion of fires and car exhausts. Carbon monoxide fits into the same wondrously designed protein pocket, and it binds to hemoglobin several hundred times better than oxygen.

So much for the best of all possible worlds and the evolutionary Plan. As François Jacob has written, Nature is a tinkerer. It has a wonderful mechanism for exploring chance variation, and, until we came along, much time on its hands. While it was banging hemoglobin into shape there wasn't much CO around. So it didn't "worry" about it.

Actually the story, of molecular evolution, is more complicated, more wondrous still. It turns out that there is always a bit of CO around in the body, a natural product of cellular processes. Heme, free of its protein, binds CO much better than hemoglobin. So the protein around the heme apparently evolved to *discourage* CO bonding a little—not enough to take care of massive doses of external CO, just enough to allow the protein to take up sufficient $O_2$ even in the presence of naturally produced CO.[19]

# ■ SEEING TO THE CENTER OF THINGS ■

We were making a television course about chemistry. There was too little money, too little time to do it right. Two teams of talented, dedicated people were at work—a group of chemists and teachers, and a television production team. It was, as I learned, an infinitely complicated process, engaging the skills and trades of thirty people, who were often at loggerheads, banging into shape a piece of video art about chemistry.

Because our budget was so low, we were dependent on "stock footage," a euphemism for free film from industrial or government sources. One tape came our way from IBM. It was about Scanning Tunneling Microscopy (STM), a marvelous technique for imaging surfaces, invented in 1982 by Gerd Binnig and Heinrich Rohrer at IBM's Zurich Research Laboratories. The technique is appealingly simple, and the images it produces so revealing that the invention was immediately recognized as of value, and rewarded by the 1986 Nobel Prize, only three years after the work was published.

To get a feel for how STM works you may read the poem that follows. The IBM propaganda, justifiably proud, turned on the hype. Now, for the first time, one could see atoms. Their tape showed a striking, false-color "fly-by" across the surface of a silicon crystal.

To the production team for our television show, daunted by the unenviable task of depicting something as intangible as the atom, the IBM footage was a godsend. They made it the centerpiece of a sequence that began with the incredibly beautiful footage of an earthrise on the moon, filmed by the Apollo astronauts, evoking the importance of the moon landing—a voyage, a search, discovery. All the directors' skill, a skill I greatly admired, shaped implicit and overt connections to the STM images. The struggle to form an atomic theory, from Democritus to Dalton, culminated in those appealing pictures of the silicon surface. Now, after such a long time waiting, now we could see atoms.

I nearly went through the roof (which was high up—we were filming in the New York Hall of Science, a structure remaining from the last New York World's Fair). What treason here to 180 years of theory and experiment that through tedious, indirect evidence built a framework of incontrovertible reality for atoms and molecules! That wondrous scanning tunneling microscope would never have been built, were it not for that painfully won, indirect but certain knowledge that atoms combine into molecules and extended arrays with precise, known geometries.

The directors ignored my concern—just another scientist adding qualifications to what anyone could see. But there was a slot, a minute long, which I was given to close the program. For that I wrote the script myself. I quickly substituted the following:

We've looked in this program at the structure of the atom. We've described the experimental and theoretical steps leading to our modern picture of the atom, this electron cloud around a nucleus. And we saw one technique, scanning tunneling microscopy, for seeing atoms. Only when we understand and see atoms, you might think, only then could we, should we, go on to the next level of complexity, to molecules.

Do you think that's true? That development of a field should await complete understanding of its foundations? I hope you don't. We knew the earth was round centuries before an Apollo astronaut took a picture of it from the moon. And we were dead sure of the existence of atoms, and knew just how atoms connected up to form molecules, before those beautiful, clear STM pictures.

Chemistry, like any human activity, proceeds simultaneously on many levels, with partial understanding, always incomplete, sometimes wrong, incredibly mostly

*Seeing to the Center of Things* (1991)

right. A great intellectual achievement of chemistry, of humanity, a chart organizing all the elements, called the Periodic Table, developed fifty years before we knew what the atom is about, is the subject of the next program.

Not only were the directors aghast at what I said, but I also said it well, with feeling. Most of the time I was not good on camera, and there was no money to send me to acting school. So I had their attention, finally. They edited out part of the offensive moon landing–STM sequence; I modified my ending. And we all understood (I think) the point that you don't have to *see* something to know, for sure, what is in this world.

*The 1986 Nobel Prize in Physics*

Because it stops
short
of touching, I feel all

the more your tongue
track the small
of my back, the hidden

line crease of leg
and buttock. You have fine
control, a feedback

loop, so that if you
touch a hair, if I rise, wanting that,
you move back, mapping

out (this is not
the first shy scan) the tense local topography.
The scanning tunneling

microscope, invented
by Binnig and Rohrer in 1982 works like this: a fine tip
of tungsten is brought

gently, mechanically
to a teasing five Ångströms of a surface.
Electrons tunnel across

the gap. Much care had
to be taken in the construction;
isolation from perturbing

vibrations being para-
mount. And control:
too close—the tip breaks,

too far—no electrons make
it across.
A sideways sweep easily

maps underlying order, local
defects,
imperfections. Sometimes atoms

jump from surface to tip, the
image shifts. On
microscopic examination the tip

is seen to be very rough. Still
the signal flows; only the asperity
closest to the surface matters.[20]

# ■ AFFINITIES ■

The chemistry of an atom is determined by its electrons. They occupy the greatest part of the atom, but, more important, they are "on the outside." When atoms approach other atoms (and that's what chemistry is about—atoms interacting with other atoms to form molecules, molecules reacting with each other, rearranging their atomic building blocks), it is the electrons which feel each other first.

Electrons may be removed from atoms, to give positive ions, for instance $Na \rightarrow Na^+ + e^-$ ($e^-$ is the symbol for an electron), or added to atoms, to give negative ions, e.g., $Cl + e^- \rightarrow Cl^-$. Such ions, surrounded by solvent molecules, are common in water or other solutions, and are especially easily formed when a salt, such as NaCl, dissolves in water.

In chemical reactions bonds are made, broken, rearranged. Atoms may be transferred from one molecule to another, as may be electrons. In general these processes proceed intimately, in the innards of the reaction.

But electrons are not like atoms. They are the carriers of electricity. They are what flows from a generator or battery, through wires, to the devices that power the world. And by figuratively and actually sticking a wire into a chemical reaction, you can watch the electron flow in a chemical reaction. This is what batteries, rusting and corrosion, and electroplating are.

*Affinities* (1990)

At left is a simple "electrochemical cell," actually half of what is needed to make a battery or an electroplating apparatus.

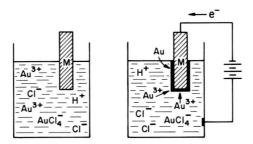

There is a container with a solution of a compound, a salt or an acid, here $HAuCl_4$. The acid is ionized in water, so that there are $Au^{3+}$, $Cl^-$, and $H^+$ ions (and $AuCl_4^-$ molecular ions) floating relatively freely in that solution. There is an "electrode," made here of the metal copper. Initially a little gold will plate out on the copper electrode. More dramatic effects will be seen if we hook two such "half-cells" together or pass an external current through the cell. Supposing we do the latter (above, *right*), in such a way as to have the copper electrode negatively charged. Then at that electrode the reaction $Au^{3+} + 3e^- \rightarrow Au$ takes place, and gold metal atoms are deposited on the electrode. Gold is electroplated on copper. Electroplating is the way (using silver ions) that silver plate is made, or that zinc is put on iron to make galvanized iron, or that tin is plated onto iron to make tin cans.

In pre-Columbian cultures gold played an essential social and spiritual role. It was symbolic of purity and royalty; its use was restricted and ritual. Inca and Aztec gold drew the conquistadors, a "compensation" for the Indies they failed to reach. A remarkable archaeological finding is that of Andean metal objects that are gilded, made primarily of copper with a thin overlay of gold. To all appearances it seems that the gold had been electroplated on the copper. Could the Andean smiths have discovered electroplating some one thousand years before Volta, Galvani, Faraday, and the other pioneers of electrochemistry?

Heather Lechtman, of the Massachusetts Institute of Technology, has solved this puzzle. No, the copper objects were not electroplated. The gold was deposited on copper by a chemical reaction that ingeniously used local chemicals to effect electron transfer without electricity.

The Andean cultures had access to the same sources of saltpeter that later made Chile a prime supplier of this important chemical. They also had native gold. To plate gold you first need to get it into solution. Gold dissolves in *aqua regia,* a venerable mixture of nitric ($HNO_3$) and hydrochloric (HCl) acids, but in little else. These acids were not available to the Andeans, but their component ions ($NO_3^-$ and $Cl^-$ and the protons, $H^+$, that make an acid strong) are readily accessible from saltpeter, common salt, and potassium aluminum sulfate (alum). Thus they apparently put together a solution effectively duplicating the "ionic atmosphere" of *aqua regia.* This brew dissolves gold, making $AuCl_4^-$ ions in solution.

It is likely that the Andean protoelectrochemists then neutralized this solution. Whereupon a copper sheet dipped into solution would prompt $Au^{3+}$ to be "reduced" (that's the chemical word for gaining electrons) at its surface to a thin layer of metallic gold. Annealing at 500–800° C would fix the thin gold layer into place. Lechtman has reproduced in the laboratory this clever sequence of ancient chemistry.

In modern electroplating the electron flow is explicit. Electrons externally supplied transform $Au^{3+}$ to Au metal. In the pre-Columbian chemistry (and there's nothing primitive about it—it's modern as can be, typical of what goes on in every laboratory today), the same electron flow occurs at the atomic level, chemically induced. Electrons revealed, electrons hidden.

The most impressive manifestation of electron transfer is a simple battery, whether it is a flashlight cell or an automobile storage battery. Batteries combine two half-cells of the type shown earlier. At one electrode electrons are given up; at the other they are accepted. A current is set up if a connection is made between the two electrodes.

Here, for instance, is the conventional "dry cell." It cannot be recharged, but it is convenient because all the components are dry pastes tightly sealed from their surroundings. One electrode is a metal-capped graphite rod, surrounded by a mixture of carbon and $MnO_2$. There $Mn^{4+}$ goes to $Mn^{3+}$, accepting electrons. So that's the positive end of the battery. The

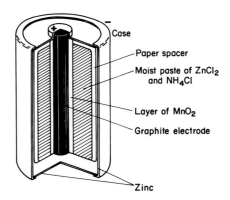

Case
Paper spacer
Moist paste of $ZnCl_2$ and $NH_4Cl$
Layer of $MnO_2$
Graphite electrode
Zinc

other end is zinc, with a paste of $ZnCl_2$, $NH_4Cl$, and water. Here electrons are generated as Zn goes to $Zn^{2+}$.

The cell runs in the direction it does because the tendency of Zn metal to lose electrons is stronger than that of $Mn^{3+}$ to do the same. Two half-reactions compete. That competition or juxtaposition is the basis of all electrochemistry, but especially of an application to corrosion that calls up another pre-Columbian connection.

Rusting is the conversion of iron to iron oxides, Fe to $Fe^{3+}$ in $Fe_2O_3$. Coating the iron with tin or paint helps, but where the coat breaks, the iron is easily oxidized. One clever idea, which you will find on ship hulls, underground gasoline tanks, or inside your water heater, is a "sacrificial anode." This is a block of magnesium (or platinum-coated titanium for ship hulls) exposed to the environment and connected to the iron vessel. The magnesium is not a random choice. This metal prefers to go to $Mg^{2+}$ more than iron wants to go to $Fe^{3+}$. So Mg controls the overall reaction. It makes the iron "want" to go from the ferric ion to the metal, the direction opposed to rust formation. The magnesium or titanium corrodes, in place of the iron.

This is anodic sacrifice of one metal to safeguard another one, one machined for some essential human use. Sacrifice of animals and humans (especially the latter in some pre-Columbian cultures) guaranteed and safeguarded the polity, a harvest, other human constructs. The analogy must not be pushed too far, but it is interesting how readily the charged word "sacrifice" comes to mind when a scientific strategy of offering up one thing instead of another is devised.

# POSSIBILITIES AND PRAGMATICS

A: The atoms in any molecule
   are held together
   by chemical bonds.

B: A line stands for a bond

A: Yes. So in $H_2O$, which is really H-O-H,
   there are two O-H bonds.
   A bond has a characteristic length,
   for O-H it's $0.98 \times 10^{-8}$ centimeters,
   and a certain strength.

B: Bonds are like springs
   if you stretch them
   or squeeze them
   the atoms bounce back

A:  What you have to do is to solve
    this equation,
    Schrödinger's equation,
    at a certain O-H distance. Then,
    repeat your calculation
    at a different geometry,
    finding this way the O-H distance
    and the H-O-H angle
    at which the energy is lowest.

B:  But what holds it together?

A:  Well, the chemical bond. You see—
    there's a stable electron configuration
    at a magic number, eight. Some atoms
    get there by taking up electrons
    to become negatively charged anions.
    Some readily yield up an electron
    to become positive cations. Anions
    and cations then attract each other.
    That's ionic bonding.

B:  Opposites attract each other

A:  Yes. Before we knew any of this
    Geoffroy in France
    and Bergman in Sweden
    made tables of affinities . . .

*Possibilities and Pragmatics* (1990)

B: And Goethe wrote a novel,
   *Elective Affinities.*
   Eduard and Charlotte
   were happily married
   until Ottilie and the Captain came

A: Yes. But getting back to oxygen, it
   attracts electrons when it binds
   with almost any element. We say
   it's very electronegative. It takes electrons
   from the two hydrogens.

B: So the bonding in water is ionic

A: Not quite. Sometimes two atoms
   coming near each other
   can reach a stable configuration,
   not by transferring electrons,
   but by sharing a pair
   (or two, or more).
   That's covalent bonding,
   common between like atoms.

B: Like attracts like

A: That's right. When we analyze
   the wave function . . .

B: What's a wave function?

A:  You'll have to take a course
     in quantum mechanics. Anyway, we find
     that the bonding in $H_2O$ is
     part covalent, part ionic.

B:  Do you mean like attracts like and
     like attracts unlike
     at the same time?

A:  Yes, I know it sounds strange.
     If only you knew quantum mechanics,
     I'd show you the wave function . . .

B:  I'm not worried. It's life-like.

# ■ CHEMICAL ARTS ■

The Romans, those master movers of fresh and waste water, built elaborate heated baths wherever they went. But soap, which began to be made in their time, they seem to have used only for medicinal purposes. To clean the body, oil was commonly used, along with mechanical cleansing agents, such as pumice and sand. Dirt was removed from clothes with fuller's earth (an absorbent clay, a hydrated aluminum silicate) and alkali—either ammonia from stale urine, or lye from wood ash, more on which later.

Soap appears to be an invention of the "less civilized" tribes to the north of Rome. The process of making it changed little until the nineteenth century. It is a chemistry as simple as it is ingenious: soap could be and was "boiled" in every household in Europe. Here's how a seventeenth-century recipe might have read:

> Mix two parts of potash (that from poplar wood is best) and one part of quicklime with water, to a thickness that an egg float on the lye. Use eight potfuls of lye for every potful of strained suet or kitchen grease. Heat the mixture to seething in a large-bottomed vessel lined with lead. If a chicken feather dissolves in the brew, the lye hasn't yet been used up. Leave in the sun for a week, stirring at intervals until a paste forms. Add musk-rose water, leave in the sun for a further week. Roll into balls, and place on waste cotton or wool in a wooden box.

*Chemical Arts* (1990)

The chemistry is simple:

$$\text{oil or fat} + \text{alkali} \rightarrow \text{soap} + \text{glycerol}$$

Soaps are salts of a base, sodium or potassium hydroxide (NaOH or KOH), and a fatty acid (a typical one is HStearate, $HOOCC_{17}H_{35}$). The fatty acid is an organic molecule, a long chain of carbon atoms and hydrogens:

HStearate ≡ Stearic Acid

Let's abbreviate the stearate group as R, a piece of chemical jargon. Then a typical soap is NaR, or sodium stearate.

Natural oils and fats are compounds called esters, in which an organic base, glycerol, combines with not one but three such long chain R groups:

triglyceride

This is a triglyceride. Soap making takes the natural oil—deer fat, whale blubber, tallow, olive oil are mixtures of triglycerides differing slightly in the number of carbon and hydrogen atoms in the R groups—and by reaction with lye makes the soap, freeing glycerol:

oil or fat      lye           soap        glycerol

In the most primitive processes, the glycerol is retained, making a soft soap. If common salt is added, the glycerol may be removed, for it dissolves in the brine, while the solid soap floats on top.

It certainly made a difference whether foul-smelling animal oils, inevitably containing the products of partial decomposition, or olive oil were used. But the sources of oil or fat are abundant. Where did one get the alkali?

In the Middle East, one had mineral sources of natron (for instance, Wadi Natron in Egypt), an important mixture of $Na_2CO_3$ and $NaHCO_3$. The corresponding potassium compounds are also present, mixed with the sodium carbonates, in the ash of any burned plant. But they are particularly abundant in the ashes of two Mediterranean shrubs, *Salsola kali* and *Salsola soda*. Here we have the sources of the systematic and English terms: *sodium* (from soda) and its symbol Na (Latin *natrium,* from *natron*), *potassium* (from potash) and its symbol K (*kalium* from *kali*), as well as the word (through Arabic) *alkali,* for the group of compounds with basic properties. *S. soda* and kelp are particularly rich in sodium carbonates; the ash of these, called *barilla* or *rochette,* was a valuable item of trade until recent times.

Water leached the carbonate from ashes, the chemistry producing a solution of sodium and hydroxide ions, lye:

$$Na_2CO_3 + H_2O \longrightarrow 2Na^+ + OH^- + HCO_3^-$$

The lye was "sharpened" by the addition of quicklime (CaO), a chemical easy to come by. The source of the alkali, actually its sodium or potassium content, was important, for the potassium soap (from potash, typical of northern Europe) was more liquid, whereas the sodium soap (from barilla and olive oil) was solid. The descriptor "Castile" attached to some soaps and shampoos is a reference to that valued product of the Iberian peninsula.

# ■ THE GRAIL ■

A young student in Kerala, India, writes of a striking color change she observes on mixing cyclohexene and iodine, a change that neither she nor her teacher understands. And having thus established her interest in chemistry, she goes on to ask: "What should I study? What in your opinion, Sir, is the most important problem in chemistry today?" The question is sincere. It cries out for a simple answer, and it hurts not to be able to give one.

In other fields the important questions seem to be breathtakingly general: How does the brain work? Let's land a manned space craft on Mars. Let's prove or disprove all of David Hilbert's conjectures left unproven. What is the "cure" for cancer? The outsider romanticizes, to be sure. But what about chemistry, where is the Holy Grail of the molecular science?

There doesn't seem to be one. Oh, occasionally some are held up for public view, gimmicky candidates for the chalice. So a few years ago, some misguided people, insensitive to ethical issues, forgetting Mary Wollstonecraft Shelley and Boris Karloff, trotted out "Our goal is to synthesize life." One can try to define other aims: to find a general anti-viral agent, to discover the catalyst for making gasoline from carbon dioxide, or to make a strong but truly biodegradable plastic. A popular textbook of organic chemistry of three decades ago cleverly decorated its backend paper with structures of molecules waiting to be made. Many of them were synthesized by the time the next edition came out. To an outsider to chemistry these seem rather limited goals.

*The Grail* (1990)

Could it be that the lack of a grandiose ambition in the science of molecules has something to do with the difficulty of arousing public interest in this central science? We don't have the very large and very distant of astronomy, with its intimations of origins. We don't have the very small of elementary particle physics. We don't have life (or we gave it away to the molecular biologists). No wonder the *New York Times* science section's coverage of chemistry is disproportionately small.

To the search for the Holy Grail, 150 of King Arthur's knights committed their hearts and resources. Translated to modern times, a like group effort demands some gigantic, or at least intricately expensive, machinery. Big science, in other words: a supercollider to search the innards, a space telescope to probe the outer fringes, a genome project to map human heredity. None of these is typical of chemistry, where from the beginning small groupings of people, working with relatively cheap cookware, have transformed a wondrous variety of matter.

Chemistry is an intermediate science. Its universe is defined not by reduction to a few elementary particles, or even the hundred or so elements, but by a reaching out to the infinities of molecules that can be synthesized. A registry of new molecules contains over ten million man-made entries. A small fraction of these is of natural origin, though millions are waiting to be analyzed. And millions more are lost in species that our ecological pressure extinguishes. Most molecules are man- and woman-made. The beauty I would claim for chemistry is that of richness and complexity, the realm of the possible. There is no end to the range of structure and function that molecules exhibit. One example is the complex three-dimensional network of Bi, Tl, Ba, Pb, Ca, Cu, and O, which makes for a great conductor of electricity. Another is a new immunosuppressant, FK-506, a large ring made up of twenty-one carbon, oxygen, and nitrogen atoms, with assorted appendages. And there is a set of new reactions conceived by a group at Merck Sharp & Dohme so as to allow them to synthesize that immunosuppressant a year after its isolation and identification from some molds.

I think that in their richness and variety molecules are to be compared with people. This is what I like about riding the subway in New York City—the incredible range of ethnicity, physiognomy, clothes, and emotions. I see tired, swarthy men, women with henna-dyed hair, people reading Korean and Russian newspapers, Caribbean blacks, a sleepy Indian girl. Angelic or rough, they're alive, and in their lives are a million novels. When I open a page of *Chemical Communications* or *Angewandte Chemie,* I get a similar feeling. I recognize the molecular types (my prejudices and education determine that), but in these pages someone has pulled off something new—here a cluster of nine nickel atoms, one inside a cube of eight, there someone else has found out the curious way an NO molecule tumbles as it jumps off a metal surface to which it had been stuck. I'm a voyeur of molecules, and—here's a difference from the world of people, as there must be—everyone wants me to look!

There is no Holy Grail in chemistry. Yes, we would like to have a magic machine that separates the most awful mixture, purifies every component to 99.94% purity (or better if we pay more), and determines the precise arrangement of atoms in space in each molecule. Yes, we'd like to know in complete detail the resistance of a molecule to every twist, bend, stretch, rock, and roll. And, yes, we certainly must espy the secret, rapid motions molecules undergo in their most intimate transformations. And, most of all, most fundamental to the science of transformations, we desire control—ways to synthesize to order, in a short time, using cheap materials, in one pot, any molecule in the world.

The secret of the Holy Grail is that it is to be found not in the consummation but in the search. Imagine every woman rejuvenated, every man saved, all ills, physical and mental, cured, all humanity perfect, and, of course, at peace. What a dull world! We've seen intimations of it, in the yoke that pseudo-socialist societies have tried to place on artists. In that redeemed world there would be no longing, no need for Swedish meatballs with fresh potatoes on a nasty day, or a consoling hug for someone inexplicably sad.

If the grand desires of chemistry were achieved—to know what one has, how things happen on the molecular scale, how to create molecules with absolute control—chemistry would simply vanish. To come to terms with complexity and the never-ending search, to find joy and beauty in the plain thing, the small step—that is the grail.

The legend of the Holy Grail was worked and reworked in the Middle Ages. In the classic version of Chrétien de Troyes, the simple Perceval first fails to cure the Fisher King because Perceval does not ask the simplest questions of "What?" and "Why?" He had been scolded earlier for asking too many foolish questions. The moral, that of science, is clear: keep asking.

Perceval was not pure enough for the more pious of the christianizers of the Celtic legend. So in later versions, for instance, in the *Queste del Saint Graal,* he is replaced by the saintlier Galahad. When the latter finds the grail at Corbenic Castle, he and his companions see a set of apparitions, a symbolic witness to the transubstantiation of the Eucharist. In these spiritual transformations is an echo of the endless changes that mark chemistry.

# ■ ON THE CRYSTAL SCALE ■

*Giving In*

At 1.4 million atmospheres
xenon, a gas, goes metallic.
Between squeezed single-bevel
diamond anvils, jagged bits
of graphite shot with a YAG
laser form spherules. No one
has seen liquid carbon. Try
to imagine that dense world
between ungiving diamonds
as the pressure mounts, and
the latticework of a salt
gives, nucleating at defects
a shift to a tighter order.
Try to see graphite boil. Try
to imagine a hand, in a press,
in a cellar in Buenos Aires,

a low-tech press, easily
turned with one hand, easily
cracking a finger in another
man's hand, the jagged bone
coming through, to be crushed
again. No. Go back, up, up
like the deep diver with
a severed line, up, quickly,
to the orderly world of ruby
and hydrogen coloring near
metallization, but you hear
the scream in the cellar, don't
you, and the diver rises too fast.[21]

*On the Crystal Scale* (1990)

## Double Carbonylation of Actinide Bis(cyclopentadienyl) Complexes. Experimental and Theoretical Aspects

**Kazuyuki Tatsumi,**[*†] **Akira Nakamura,**[†] **Peter Hofmann,**[*‡] **Roald Hoffmann,**[*§] **Kenneth G. Moloy,**[∥] **and Tobin J. Marks**[*∥]

*Contribution from the Department of Macromolecular Science, Faculty of Science, Osaka University, Toyonaka, Osaka 560, Japan; Anorganisch-Chemisches Institut der Technischen Universität München, D-8046 Garching, FRG; Department of Chemistry, Cornell University, Ithaca, New York, 14853, USA; and Department of Chemistry, Northwestern University, Evanston, Illinois, 60201, USA. Received January 13, 1986.*

. . . *Materials and Methods.* All procedures were performed in Schlenk-type glassware interfaced to a high-vacuum($10^{-5}$ torr) line or in a nitrogen-filled Vacuum Atmospheres glovebox equipped with an efficient, recirculating atmosphere purification system. Argon (Matheson, prepurified) and carbon monoxide (Matheson, prepurified) were further purified by passage through a supported MnO oxygen-removal column and a Davidson 4-Å molecular sieve column. Pentane ($H_2SO_4$-washed), heptane ($H_2SO_4$-washed), toluene, and diethyl ether (all previously distilled from Na/K/benzophenone) were condensed and stored in vacuo over $[Ti(\eta^5\text{-}C_5H_5)_2Cl]_2ZnCl_2$ in bulbs on the vacuum line.

*Steps and Processes* (1989)

$^{13}CH_3I$ (99% $^{13}C$, Cambridge Isotope Laboratories) was degassed by freeze-pump-thaw cycles on a high-vacuum line and dried by condensing in vacuo onto $P_2O_5$ and stirring overnight. $^{13}CH_3Li\text{-}LiI$ (99% $^{13}C$) was prepared in the conventional fashion by reaction of $^{13}CH_3I$ with washed Li sand (1% sodium, 30% dispersion in mineral oil, Alfa) in diethyl ether. The complexes $Cp_2^*ThCl_2$ and $Cp_2^*Th(CH_3)_2$ were prepared by our published procedures. The complex $Cp_2^*Th(^{13}CH_3)_2$ (99% $^{13}C$) was prepared analogously to $Cp_2^*Th(CH_3)_2$ employing $^{13}CH_3Li\text{-}LiI$ (99% $^{13}C$). Chemical and isotopic purities were checked by H NMR and infrared spectroscopies and by mass spectrometry.

*Physical and Analytical Measurements.* Proton and carbon NMR spectra were recorded on either a Varian EM-390 (CW,90-MHz) a JEOL FX-90Q (FT,90-MHz), or a JEOL FX-270 (FT,270-MHz) spectrometer. Infrared spectra were recorded on a Perkin-Elmer 599B spectrophotometer using either Nujol or Fluorolube mulls sandwiched between KBr plates in an o-ring sealed, air-tight holder. GC studies utilized an 8 ft 5% FFAP on Chromasorb G column (isothermal operation mode at 110°). Mass spectra were recorded on a Hewlett-Packard Model 5985 GC/MS with interfaced data system. Solids were studied by the direct injection technique. We thank Dr. Doris Hung for assistance with these measurements.

*Crossover Experiments.* In a 25-mL reaction flask was placed 0.050g (0.094 mmol) each of $Cp_2^*Th(^{12}CH_3)_2$ and $Cp_2^*Th(^{13}CH_3)_2$. The vessel was evacuated, and then 10 mL of $Et_2O$ was condensed into the flask at $-78°C$. The suspension was stirred at this temperature until all of the material had dissolved and a colorless solution was obtained. The flask was then backfilled with 1 atm of CO and the solution stirred vigorously. After 4 h at $-78°C$, the solution was allowed to warm to room temperature whereupon a colorless solid ($[Cp_2^*Th(\mu\text{-}O_2C_2\text{-}(CH_3)_2)]_2$) precipitated. Next, 2 mL of degassed 1 M $H_2SO_4$ was added to the reaction mixture via syringe under a flush of argon. After the resulting suspension was stirred for 15 min, the mixture was centrifuged to remove a colorless, flocculent solid. The $Et_2O$ layer was then separated from the aqueous phase. The aqueous phase was next washed with four 2-mL portions of $Et_2O$, and the washings were combined and then dried over $MgSO_4$. The extracts were then concentrated to ca. 1 mL on a rotary evaporator and analyzed by GC and GC/MS.

Besides solvent, the only organic products detected under these reaction conditions were pentamethylcyclopentadiene and 3-hydroxy-2-butanone, the latter in isotopic ratios, 100(2) $^{12}C,^{12}C$:3 (2) $^{12}C,^{13}C$:98 (2) $^{13}C,^{13}C$. These results were additionally calibrated by control hydrolysis experiments with individual samples of $[Cp_2^*Th(\mu-O_2C_2(CH_3)_2)]_2$, $[Cp_2^*Th(\mu-O_2C_2-(^{13}CH_3)_2]_2$, and Aldrich 2,3,5,6-tetramethyl-1,4-dioxane-2,5-diol, which dissociates to 3-hydroxy-2 butanone in the 150°C injection port of the GC. . . . [22]

# ■ F O R C E S   C O N S T A N T ■

Chemists distinguish three kinds of motion of microscopic or macroscopic bodies: translation, rotation, and vibration. They are best illustrated by an example. Consider a molecule made of three atoms in a line.

●——●——●          O＝C＝O

This could be carbon dioxide, $CO_2$. In "translation," the tiny molecule moves as a whole, in any direction. Here are two possible translations, indicated by little arrows showing the tiny displacements of the atoms:

In "rotation," as the name implies, the molecule rotates as a whole:

*Forces Constant* (1991)

In "vibration," one piece of the molecule moves one way, one another. Restoring forces are at work, like little springs, keeping the molecule in a certain favored geometry. So, after starting out vibrating one way, the atoms eventually reverse the motion:

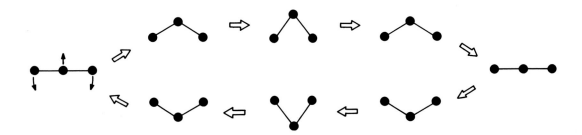

Real molecules, carbon dioxide in our lungs, for instance, are doing all of these motions simultaneously. It's a wild dance floor there at the molecular level.

By irradiating the molecules with light we can actually trace the details of molecular motion. What we are doing then is putting energy, from the light, into molecular motions. The motions turn out to be quantized, which means that they can take up energy only in certain specific doses. If a vibration is difficult, that is, the springs resist bending more, it will take more energy, light of greater frequency, shorter wave length (to the ultraviolet) to get the molecules vibrating. By scanning the energy absorbed from light (this is called spectroscopy), we can map out the energies required to effect the different motions.

When a molecule falls apart into two pieces (that's one simple chemical reaction), the energy the molecule initially possesses may show up in different ways in the products. Take our triatomic A-B-A molecule and consider two ways it could fall apart into A-B and A:

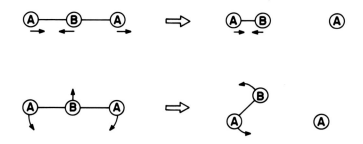

In the first process, molecules A-B and A move apart (translation) and A-B, one product, may also vibrate, excited by the process of decomposition. In the second alternative the product molecules separate, but also A-B is set tumbling, rotating relative to the original center of mass.

In experiments, using laser spectroscopy, what one actually does is to determine the distribution of energy between translation, vibration, and rotation in the product (A-B + A). From that one reasons back to the way the reaction proceeded. Which is fun to know.

# ■ PATTERNS ■

Mark Kac, a mathematician with great physical insight, once wrote an article called "Can One Hear the Shape of a Drum?" He showed that if you could listen perfectly, hear every tone, harmonic, and overtone, you could deduce from all these frequencies the area of the drum's vibrating surface, its perimeter, and the number of holes in it—without seeing the drum!

Knowing without seeing is at the heart of chemistry. Matter is recalcitrant—green, slimy, nutritious—but what is in it? When it became clear that we must think of atoms and molecules as universal building blocks, the problem that they were tiny remained. No nineteenth-century microscope could resolve molecules; magnification by a factor of 1,000 got nowhere near the molecule. Still we learned to know them, very well.

To see how this was done, let's pursue the musical instrument analogy a step further. The fundamental frequency of a guitar string is a function of its length; the shorter the string, the higher the note. The pitch of the tone also depends on the string's thickness and density, often determined by the material from which it is made. You can tell one instrument from another playing the same note because of its timbre, the mix of overtones and harmonics into the fundamental note. If you have a good ear, you can recognize the instrument, and, if it is a guitar, with some science deduce the length of the vibrating string. You can't do it if the instrument is quiet. The string needs to be twanged.

*Patterns* (1991)

Molecules don't seem like musical instruments, but in a way they are. The water molecule is an assembly of two hydrogens and an oxygen, bent, H-O-H angle 104.5°, with certain definite O-H distances. Hydrogen and oxygen are heavy. A reasonable model for the molecule is one of masses linked by springs (those are the chemical bonds). The molecule vibrates, and it rotates, set into motion by its many collisions with other molecules. Water's electrons also have available motions. In fact, this essential molecule, were it possible to observe it directly, would outdo any kinetic sculpture.

All these motions turn out to be quantized, to be allowed only at certain, definite energies. For each motion there is a ground, nearly resting state as well as higher, excited states, specific dollops of energy above the ground state.

Now we twang the water molecule, gently, just by shining light on it, light of all colors. Actually, the light that would make water dance is not visible but in the infrared range. The water molecule ignores all colors, except that whose energy matches what it needs. (Einstein derived a relationship between the energy of a photon and its wave length, $E = hc/\lambda$. Here c is the speed of light, $\lambda$ the wave length, h a fundamental constant.) That light it absorbs.

This is the essence of a spectroscopic measurement. A typical spectrum is shown below.[23] Light shines on an object; where its frequencies match the characteristic tones and overtones, light is absorbed. The instrument compares light intensity coming out to that put in. The valleys are where the molecule took in the energy of light, turned it into its motions.

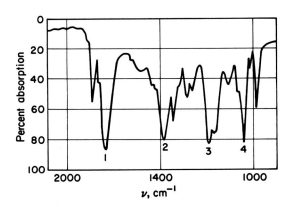

Infrared absorption spectrum of vitamin C (ascorbic acid). Chemists often use the reciprocal of the wavelength ($cm^{-1}$) for measuring infrared absorption peaks. The strong absorptions at **1** (1640 $cm^{-1}$), **2** (1316 $cm^{-1}$), **3** (1136 $cm^{-1}$), and **4** (1042 $cm^{-1}$) help to identify vitamin C.

A different shape, stronger or weaker bonds, a different molecule—all of these would lead to absorption of different colors—a different timbre. With a theory and sufficient detective work, the molecules speak to us. In their wondrous quantized motions they sing a song of identity, tell us who and what they are in a tune that our instruments, now our measuring instruments, may hear.

# ■ FORMULATION ■

The top ten hits of the chemical world in 1991 are:

1. Sulfuric acid
2. Nitrogen
3. Ethylene
4. Oxygen
5. Ammonia
6. Lime
7. Phosphoric acid
8. Sodium hydroxide
9. Chlorine
10. Propylene

These chemicals are arranged in decreasing order of weight produced in the United States. First, leading the charts for over a hundred years, is $H_2SO_4$, sulfuric acid. More than 87 billion pounds of it were synthesized in 1991, and I assure you this quantity was not made for fun. It was made because someone out there was willing to buy it.

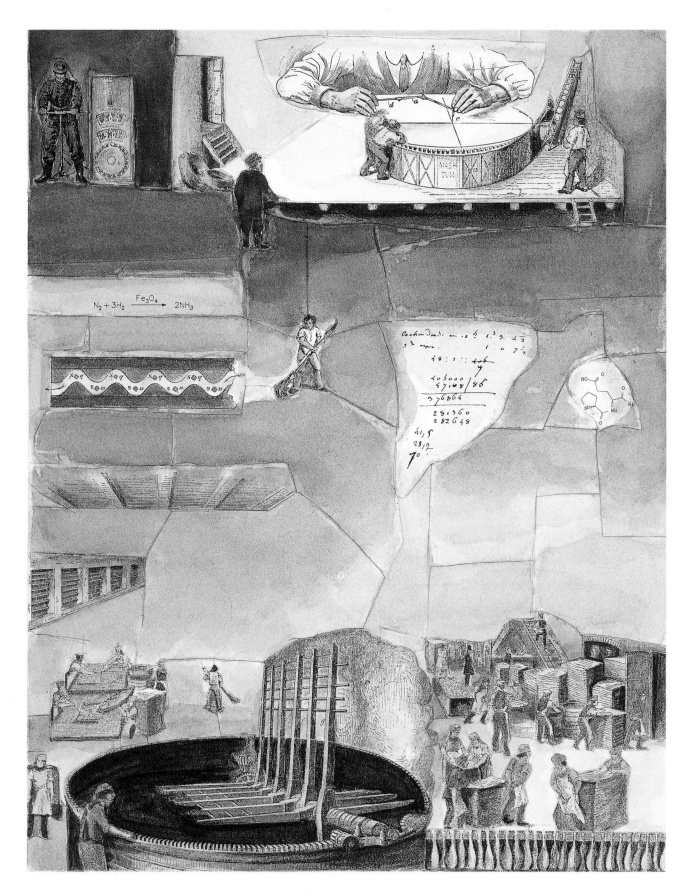

*Formulation* (1991)

Where does all this incredible volume of corrosive acid go? Into just about every chemical process: into petroleum refining, metallurgy, paint, polymer, paper, detergent manufacture, and, most important (along with lime, ammonia and phosphoric acid in the top ten), into chemical fertilizers. Sulfuric acid is the ultimate feedstock, go-between, transformer. It changes matter, and is changed itself.

$H_2SO_4$ is responsible for most of the crucial sulfur in our amino acids. Combined with phosphate rock, the corrosive acid becomes superphosphate fertilizer, not very tasty but hardly harmful. It is then taken up by wheat, where its sulfur atoms are incorporated into crucial amino acids. I eat a bagel, and the sulfur originating in the synthetic fertilizer is transformed further, into my proteins.

Should you worry about your bagel falling apart into sulfuric acid and phosphate? No, the chemical transformations are as irreversible in our lifetime as your getting older, as a steak being grilled, as cement setting. Please don't worry.

You can look at the sulfuric acid production of an industrialized country and extrapolate from it to that country's Gross National Product. This is because sulfuric acid is involved in nearly every industrial process.

The top ten list changes slowly. In 1991 no new chemical entered the list, none dropped out of it. But over a longer time interval, say fifty years, there is change. The new kids on the block since 1940 are ethylene and propylene, the raw materials of the polymer century, the source of the vast majority of synthetic plastics and fibers.

# ■ NATURAL CYCLES ■

In this century science and technology have transformed the world. What we have added, mostly for the best of reasons, is in danger of modifying qualitatively the great cycles of the planet. We see the effects of our intervention in the change in the ozone layer, the pollution and acidity of our waters, in why we wash an apple, in the crumbling statuary, our heritage, dissolving.

The effect of science and technology was surely felt before. But not till this century did the man- and woman-made, the synthetic, the unnatural, truly contend with nature. Is this a time to praise, a time to fear?

The world that men and women entered before there ever was such a thing as chemistry was not a romantic paradise but a brutish, inimical environment in which men and women hardly lived past forty. That natural world was transformed by our social institutions, our art, our science. Certainly not by science alone. We do not kill female children, nor keep slaves, nor let the sick die, all practices some societies, I'm sorry to say some religions, once thought natural. Even though we have such a long way to go, we have changed our nature. Our lives are improved by detergents and synthetic fibers, *and* by a social web of human, constructed support. Our lives are enriched by Mozart and Bob Marley and the Wailers, bringing to us a world of synthesized, transformed beauty and satisfaction.

Yet we also use our transforming capacity destructively—to annihilate a quarter of the species in this world, to hurt our brothers and sisters. It is we who do this; there is no hiding behind a "they." This seems to be our dark side. We have a problem in finding a balance, with not letting our transforming nature run amok; we seem to have difficulty in *cooperating* with our own world.

In the tradition I come from, the Jewish tradition, there is a concept that is relevant to this theme of natural/unnatural. It is *tikkun*. The word literally means "repair"—of a shoe, but also of a soul, of the world. The sense is of change by human intervention. So the word's meaning shades over to transformation. *Tikkun olam*—the transformation of the world, by human beings, more than a salvaging, a making of our future consistent with what we are given.

Friends, it is not given to us *not* to make new things—be they molecules, a sculpture, or a civil rights bill that a president vetoes. We are sentenced by our nature to create. But we do have a choice, to fashion this world in consonance with the best in us, or the worst. One can doubt about whether our transformations are of human value. But there can be no doubt as to what they *should* be.

*Natural Cycles* (1991)

It is impossible to fill all of normal three-dimensional space with a tight packing of ideal tetrahedra. Attempts to do so lead to frustration, little left-over nooks and crannies. David Nelson, a physicist at Harvard whom I once taught chemistry, is an expert at gauging the frustration of space due to its unfulfilled longing for tetrahedral filling. One day he showed me this beautiful structure:

You see a one-dimensional chain of tetrahedra. Each tetrahedron shares one face with the next one along the chain. Alternatively, you see a gently winding triple helix of points, or a propagating staggering of triangles.

Face-sharing tetrahedra are an extremely rare motif in the otherwise structurally rich world of inorganic molecules. There do exist discrete molecules in which two or three tetrahedra share a face. The tetrahedra may be empty, as in osmium clusters covered by carbon monoxides made in Cambridge, England, by Brian Johnson, Jack Lewis, and their coworkers. Or they may be filled by metal atoms, as in some copper iodides synthesized by Hans Hartl in

*Celebrating Solutions* (1991)

Berlin and Susan Jagner in Gothenburg. But three seems to be the limit; no one has yet made a chain of four or more face-sharing tetrahedra.

Here then is an opportunity for theorists, usually relegated to producing *ex post facto* rationalizations of the incredibly beautiful products of chemists' minds and hands. Could it be that for a certain metal at the center of the tetrahedron, for certain atoms at the corners, that such a helix might be stable?

Theory comes in, for eyeballing the structure is not of itself instructive. A glance at your daily newspaper will remind you that not all that is beautiful is attainable. From a recently published sequence of quantum chemical calculations, Chong Zheng, David Nelson, and I reason that the helix has the best chance of materializing for WI, tetrahedra of iodine centered by tungsten. Or for PtCO, empty tetrahedra of platinum, with carbon monoxides radiating out around the helical core.

It turns out that we didn't think up this appealing structure. First, it appears as a problem in H. M. S. Coxeter's influential geometry book. Second, a member of my research group spotted it as an outdoor sculpture by Ted Bieler, standing in a public space in Toronto. Arata Isozaki, a leading Japanese architect, has the helix rising several hundred feet above a building in Mito. Finally, this geometry, dubbed a "tetrahelix," is the centerpiece of a chapter of Buckminster Fuller's *Synergetics,* published in 1975.

Given this provenance, the molecular structure can hardly resist materialization. If artists, mathematicians, and architects can think up a tetrahelix, even build one, could synthetic chemists tolerate being so far behind?

# ■ BETWEEN WONDER AND PROOF ■

## *The Collages of Vivian Torrence*

### LEA ROSSON DELONG

Wonder and intuition are central in the work of Vivian Torrence. So it seems appropriate that she works in collage, a medium introduced about 1912 by Picasso and Braque, and most notably used by surrealist artists such as Max Ernst and Joseph Cornell. Collage uses pieces of paper cut from disparate sources and fits them together to create a new image with a new meaning. Collage breaks up the common path of thinking as a logical, commonsense process and takes it to a realm where relationships and conclusions are not so clear-cut and familiar. Two images placed side by side may have no apparent connection. Understanding the relationship intended by the artist requires an exercise in intuition or imagination. Such works of art demand that we use both our logical and nonlogical faculties.

For example, in the collage *Radium* (p. 31), Torrence deals with Marie Curie's discovery of that radioactive element. Instead of painting a single incident from the scientist's life, Torrence uses a series of forms that seem at first unrelated. Further reflection reveals that these forms suggest a range of ideas related to Curie: her method of working, her family, and how her work affected her physically. Even more challenging, Torrence comments on the significance of the discovery and its impact on the future. How could all these subjects be sensibly integrated into a single narrative? In a conventional way, it seems unlikely that they could be. But through collage Torrence can choose a single image that can stand for or evoke a larger set of ideas. The whole composition then creates a web of ideas that add up to a multi-layered comment on the subject at hand. The links among the individual images within the collage are not always easy or obvious; sometimes they require a fairly high level of knowledge (to identify, say, a Bohr model of an atom). But most of the time a person with no specific knowledge of chemistry can get an idea of the meaning. More careful analysis will yield a more richly detailed understanding, but the collages are accessible and rewarding on a variety of levels.

In *Radium* we see a portrait of Madame Curie, a portrait that appears slightly fuzzy, as if the figure had been irradiated. Across from it there is an X-ray of a hand. The wedge-shaped form across the fourth finger is a wedding ring, indicating that Curie was also a wife and mother at the time she conducted her early experiments into radioactivity. Although her husband, Pierre Curie, is not shown in the collage, the wedding ring is a clear reference to the partnership, both scientific and marital, of the two scientists.

Directly below the portrait is a woman working with some sort of apparatus, symbolizing Mme. Curie's long hours in the laboratory. Across from this is a mass of dark, undifferentiated matter that represents the tons of pitchblende from which the Curies extracted their tenth of a gram of radium. Looking at this daunting, uninviting mass gives us some sense of the enormity—even the physical obstacles–of the Curies' task and of their achievement as well. The hand and energy graph above the pile of pitchblende remind us that Mme. Curie suffered radiation burns on her hands because of direct contact with radioactive material in her research.

The upper section of the collage deals with the issue of Curie's accomplishment. Between her portrait and the X-ray of her hand is a model of a radium atom (devised by the Danish physicist Niels Bohr [1885–1962]), and to the left of the atomic model are hands using a carving tool, a symbol of God carving and shaping the universe. Whether we see them in exactly these terms or not, the idea of constructing or shaping something is usually seen as a positive action. Across from this hand and on the other side of the atom is a dark-suited medieval knight who symbolizes the dreaded consequences of our knowledge of such atoms. On either side of the atom are our choices: the hands of construction or the dark knight of destruction. The knight might also be a reference to the fact that Mme. Curie herself died of anemia, caused by her prolonged exposure to radioactivity.

These are some of the ideas that arise from the symbols in the collage. The images, snipped from different sources, don't add together into a seamless, flowing narrative. There are gaps that can be bridged only by our own imagination or intuitions about the world. Torrence's collages require that we, the observers, make the links that help create an experience of insight or intellectual satisfaction. When confronted with these collages, perhaps we are somewhat like the scientist who has this fact and that fact but no apparent relationship among the findings in his or her possession. It is at this point that one must stretch a sense of what is possible and employ some kind of creativity to guess at what the answer or resolution might be.

Torrence's use of collage, an established tradition in twentieth-century styles of fantasy and surrealism, is not surprising. But a second aspect of her work, an interest in science, is perhaps less expected. Although there may be more connections between twentieth-century art and science than we generally assume, an artist who works so consistently with scientific themes is unusual.

To Torrence, the compelling aspect of science, chemistry in particular, is its concern with basic, elemental questions. She is interested in the kind of questions scientists ask, especially theoretical scientists: How do things begin? What makes them continue or cease? What meaning should humans derive from the phenomena, on all scales, of nature? Such questions, Torrence might say, were answered in the past mostly by theology, with artists frequently asked to give form to those answers. For instance, in the Sistine Chapel ceiling, Michelangelo pictured God giving solid form to a whirling, chaotic mass as he creates the world. Michelangelo's painting is not a scientific illustration, yet it does certainly reflect the common answer of his time to the questions of how the world began and where matter originated.

Whether we turn to theology, science, or art for answers, these are questions that all people of thought must ask at some point. As an artist, Torrence deals with fundamental issues of humanity and our relationship to nature, and she understands that the chemist too is continually working with these kinds of issues. "It all begins with wonder," she has written to explain her motivation for the collages of the Chemistry Imagined series. Part of the goal of this series was to retain that quality of wonder and mystery and yet to work with the specific subject matter of chemistry, its history, its practice, and its aspirations.

Ten of the collages have to do with the history of chemistry, four in particular with the ancient Greek idea that matter could be divided into four basic elements: air, earth, fire, and water. These collages, the first produced for the series, provide an introduction and guide to the remaining works and should be looked at in some detail, as we find in them recurrent characteristics and themes.

Each of the four employs line-engraved images originally found in nineteenth-century books (printed before photography was widely used for illustration). They incorporate mythology, especially the effects the gods had on the everyday affairs of humans. They address the issue of how knowledge is gained or imparted and, more fundamentally, how humans are excited to investigate their world in the first place. The human aspiration to understand and know more—and perhaps have some measure of control—are ideas that recur. Each of the four collages suggests some notion of perfection, of an absolute knowledge, and of the gods or the forces that move beyond our sight and, perhaps, beyond our comprehension. Finally, we are sometimes reminded of the ecological situation of the earth today.

In *Greek Air* (p. 47), Hera, wife of Zeus and queen of heaven and earth, looks dispassionately at a steaming planet Earth surrounded by swirling clouds. Above her head, like a halo, is a Bohr model of a hydrogen atom. A distillation column conducts heat generated by the earth to a group of agitated, panicked humans in a too-hot tub. Their excited, disorderly state reminds us of the Greek word that survives in our "chaos." If we heat up our earth too much, the collage seems to caution, it will produce our own chaos.

Into this situation, and unnoticed by the preoccupied humans, enters a figure who is part animal, part human. Might he rescue the humans or tell us how to change things for the better? Figures such as this, who act as a link between the gods and humans, can be found throughout the Chemistry Imagined series. This demi-god is a symbol of inspiration for humanity. Where does the inspiration to examine and affect our world come from? This is a question the artist asks throughout the series, not only for herself but on behalf of chemists and others of us as well.

In *Greek Earth* (p. 51), Torrence makes the most direct reference to art and the work of the artist in transforming substances. The solidity of earth, the most stable of the Greek elements, is symbolized by the geometric forms that dominate the collage, especially the square. The overall composition has a somewhat geometric appearance and gives subtle form to that idea of stability, which is particularly apparent when compared with the flowing, circular composition of *Greek Air*.

The narrative of *Greek Earth* begins at the top of the composition with the exactly ordered atomic models and perfect sphere. This sphere is then shown in the hands of God as He takes a tool and breaks up its order and perfection. Through a geometrically shaped barrier (the square slab with the circular hole in the middle), He throws down to earth the disorganized mass which, as it falls and breaks up, looks something like a meteor. That meteor symbolizes partly the undifferentiated matter that forms the earth itself. Tiny figures on a mountaintop to the right gaze without any apparent notion of the source of this heavenly display. How could they, after all, since the workings of God, the first forces, are hidden from them by the shield? Such shields or barriers are found in several forms throughout the series and always serve to show the separation of one realm from another: heaven from earth, perfection from disorder, knowledge from ignorance, aspiration from realization—whatever separates humans from God or from harmony, order, or knowledge.

Closer to the foreground, an untutored worker oblivious to the events above him struggles to dislodge a stone. Similar, though smaller, stones reappear on the cube, and a classically garbed female figure (symbolizing the muse of sculpture) stands nearby. These images might call to mind stories of sculptors searching the quarries for the right stone from which to carve their figures. In blessing these raw materials from the earth, the muse provides the inspiration that results, on the left, in sublime and beautiful works of art such as the Venus de Milo. Such works give no hint of the ungainly common rock from which they were taken. According to certain aesthetic theories, when a work of such order and beauty comes into existence, we regain a glimpse of the perfection from which it sprang long ago. The material has gone through many manifestations and mixtures, but now through the work of art it has reclaimed the unity and purity it once possessed in the hands of God. This cycle is expressed by a circular symbol for unity that hovers over the head of Venus on its way back to the heavenly source.

Theories of beauty, what its characteristics are and how we have such a concept in our minds to begin with, are abundant and complex during the Renaissance and other times. Although Torrence does not illustrate any particular theory in her work, recognizing the tradition of these interpretations of beauty can help us understand the collages.

The remaining two collages from the Greek Elements group continue the motifs of unperceived gods as the source of power. In *Greek Water* (p. 59), Zeus positions himself at the crest of Niagara Falls, and in *Greek Fire* (p. 55) he dispenses not only fire itself but the understanding of fire, especially of how to control it. The god's allowance of knowledge to humans is sporadic and capricious, as symbolized by the shower of numbers he tosses out in *Greek Fire*. Both collages include the demi-god or go-between who takes some interest in human affairs. In *Greek Water* the Cupid at Zeus's shoulder points to the swimmer, who symbolizes humans trying to make their way through the watery substance as they aspire to the knowledge of the gods. The swimmer directs himself toward the falls and the source of power. How close he will come to that source is something the collage does not reveal.

Several of the collages have to do with the *wrong* answers. In chemistry, as in most other disciplines, it's the right answers we want to talk about, not the mistakes or scientific blind alleys. Wrong answers are discussed reluctantly and then only to trumpet the achievements of a more enlightened age that has cast off the burden of entire systems of thought that proved false. But Torrence has a particular interest in these "wrong" answers. Part of their appeal is the complex systems of procedures, symbols, and literature that are built around ideas that were fallacious. Alchemy is the most famous chemical example. Its roots are deep in Antiquity and it was indisputably a progenitor of modern chemistry, yet its philosophical core was long ago repudiated. Certainly, alchemy contributed an elementary knowledge of the nature of some chemicals and some reactions and processes. Yet at its base was the magical and erroneous belief that ordinary substances, such as lead, could be changed into gold. Common, lowly materials could become that precious rarity if the right processes and materials were used. Even further, alchemy proposed that the right combination could produce an elixir of eternal life and that the person who possessed this knowledge could be transformed from an ordinary, flawed human being into a person of purity and godliness. In alchemy the quests for human perfection and control over nature were interwoven.

Chemicals were personified and given human or celestial traits. Some alchemical formulas, for example, gave male or female characters to substances, proposing that their reactions produced a blend of the two that alchemists symbolized as the "holy hermaphrodite." These ideas were represented by cryptic symbols that were indecipherable to all but the alchemist. Such symbols and personifications, quests for mysterious formulas that could transform common substances into rare ones, and hopes of finding human perfection through the manipulation of chemicals are no longer part of chemistry. Such associations are so discredited, in fact, that it may be embarrassing for present-day scientists to admit they ever existed. Hoffmann's essay addresses some of the ambiguities in modern-day chemists' attitudes toward alchemy.

The theory of phlogiston is nearly as discomfiting as alchemy. In the eighteenth century the idea that the substance phlogiston was responsible in one way or another for chemical reactions involving burning answered many questions. It was a firmly entrenched, widely accepted "truth" of science then, and a complex system of research and information evolved around it. But it was wrong. Intellectually the concept was satisfying, and there was an internal logic to all the experimentation and literature based upon it. One of the enduring themes of Torrence's work has been situations like these: ideas or explanations that are carefully thought out, beautifully ordered, and intellectually balanced and satisfying, but wrong.

In *The Philosopher's Stone* (p. 79), Torrence looks at alchemy not just as a chemical version of the Dark Ages but as a system of thought. It was thoroughly laced, of course, with magical, occult elements, but at its foundation was a very serious quest—one that endures today. One of the mainstays of the alchemical quest was the Philosopher's Stone—the catalyst that could change ordinary material into gold. The search for the right combination of substances that could trigger this conversion produced many formulations whose main accomplishment was to advance the understanding of chemical behavior. But as for changing lead into gold or refining the human soul, no true success was ever achieved. Still, it is the belief and the resolute search through the ages that fascinate the artist.

Other collages that deal with more contemporary chemical matters, such as electron transfer and electron sharing, make it clear that our world is not static—things seem to be moving quite a bit, whether we witness it or not. Even the atoms of substances we consider lifeless or inorganic are engaging in motion of some sort internally. This ever-present movement is the subject of *Forces Constant* (p. 141). We have often been told that matter consists mostly of empty space. But like the space in an analytic cubist painting, its only empty sometimes. The electrons that surround the nucleus move constantly, and so parts of the empty space of an atom are occupied for at least a little while. As Hoffmann's essay implies, movement is not a simple, going-around-in-a-circle situation, but is complicated, full of variations in the shapes of the orbits or the angles of alignment. As he demonstrates, molecules join up and molecules fall apart. Determining what exactly their movement is and what provokes or controls it is an exacting study in chemistry. But for the average person, just having an elemental glimpse of the amount and variety of motion in matter is astounding. The lack of stasis is almost shocking. It is truly as Hoffmann says "a wild dance floor there at the molecular level."

Yes, but what is also amazing is that things don't fly apart any more than they do. In Torrence's collage, we see groups of wrestlers struggling with one another. They pull this way and that, succumb to one grip, then slip out of another. Above them a child balances on a wiry lattice structure, while a fencer poised atop a molecular model points to a molecule. Amid all these vigorous, exercising figures, some kind of balance seems to be maintained—no one seems to be winning or losing. In the airy regions above all these contending forces, models and drawings translate their motion to the molecular level where, we assume, a similar equilibrium is attained.

Just as motion is a challenge to the scientist, so it is to the artist. It's always a problem to depict motion on the still surface of a canvas, though people like Watteau and Degas produced some remarkable successes. Yet equilibrium is difficult as well. Equilibrium isn't just being

165

still, it's bringing "forces constant" into some sort of balance where neither force loses its special character. One can't win, but the other can't lose either. The twentieth-century artist who has dealt with this issue most directly is Piet Mondrian (1872–1944). The Dutch artist abandoned representation in the 1910s mainly because he believed material objects couldn't carry the ideas he wanted to express. He believed (in contrast to the Dadaists) that there was order and regularity in the universe but that naturalism in art could never reveal it to us. We could know and express that universal order only through our intellect or our intuition. So he used only the most essential, fundamental forms available to art: primary colors and straight lines (either vertical or horizontal). To demonstrate universal harmony through art, the straight lines (never free-form, curved, or diagonal) met at right angles only: 90°—nothing more or less. At this right angle, or the juncture of the two fundamental linear forces of horizontal and vertical, equilibrium was established. Neither line was robbed of its energy. Both were brought into balance. With thoughtful looking at a Mondrian painting, you can almost hear the humming of lines still asserting their basic energy but still kept in equilibrium.

Contemporary artists such as Torrence have inherited Mondrian's devotion to fundamental issues and their expression in art. Unlike Mondrian (and affected by Dada), she chooses not to work with abstraction. Instead, she uses pre-existing images that require the exercise of intuition to understand. Our knowledge and experience of the natural world is not rejected, but it still isn't enough for understanding. To come closer to that, we need to rediscover our intuition and creativity and bring them into play.

As annoying as it might be to have artists (or scientists) who insist on pushing us farther than we care to be pushed, it is still to our advantage to have them among us. Mondrian understood well the position of the artist (or anyone) who deals with things many of us would just as soon leave alone. In 1937 he wrote: "The pioneers . . . discover consciously or unconsciously the fundamental laws hidden in reality, and aim at realizing them. In this way they further human development. They know that humanity is not served by making art comprehensible to everybody; to try this is to attempt the impossible. One serves mankind by enlightening it."[24]

Throughout the Chemistry Imagined series, chemistry is seldom treated as the science that brings us a better detergent or some other improvement in the products of daily life. It is presented as a science primarily of intellect and imagination. Its theoretical and creative aspects are emphasized rather than its practical applications. Even when pragmatic issues are introduced, it is usually implied that their ultimate source is the more purely creative realm of thinking. It seems that art too can be divided into the more pragmatic, down-to-earth kind and one that deals with a higher level, both in source and inspiration. Art can be a frustrating and disappointing experience for those who require that the artist simply report the everyday world. It might be helpful to think of art as a form of inquiry or investigation into things. Any sort of human experience—imaginative, intellectual, or material—can be the subject of art. Asking that art restrict itself to simple representation is like asking chemistry to stay within the bounds of industrial production. Not only would it make both disciplines less interesting, it would rob them of the curiosity and daring that keep them vital.

The chemist's role and hope, Torrence explains is "to find out what *stuff* is," or to examine, analyze, and then, most important, think about it in an unconventional, creative way. In her collage *The Chemist,* she places a God-like hand above the chemist, a hand that not only creates the "stuff" but offers the inspiration to understand it. The chemist constructs a world through models and equations. How closely do they correspond to the actual appearance of things? It doesn't matter since the point of the chemist's work is not to show us what things look like but rather how and why they exist and work the way they do. In discussing this collage, Torrence speaks of it as having to do with "mental constructs of a more perfect form than the original stuff." Perhaps in truly comprehending our world and our existence in it, it is important to look beyond external appearance and, as Mondrian directed, discern an underlying beauty and purpose through our creativity and intuition.

# ■ N O T E S ■

1. Drawing of the NaCl structure reproduced by permission from K. W. Whitten, K. D. Gailey, and R. E. Davis, *General Chemistry with Qualitative Analysis,* 4th ed. (Philadelphia: Saunders, 1992), 260.

2. The β-D-glucose structure was determined by W. G. Ferrier, *Acta Crystallographica* 16 (1963): 1023.

3. The representation of water was provided by David L. Beveridge, Wesleyan University, whom we thank.

4. For an extended discussion of representation in chemistry, see R. Hoffmann and P. Laszlo, *Angewandte Chemie, Intern. Edit. English* 30 (1991): 1–16.

5. We are grateful to Donald B. Boyd, Eli Lilly Co., for providing structures **4–7**.

6. Photographs by Charles D. Winters, SUNY-Oneonta; reproduced by permission from J. C. Kotz and K. F. Purcell, *Chemistry & Chemical Reactivity,* 2d ed. (Philadelphia: Saunders, 1990), 106.

7. Quotation by Sture Forsén from T. Frangsmyr, ed., *Les Prix Nobel 1986* (Stockholm: Almqvist & Wiksell, 1987), 23.

8. Quotation is from E. Curie, *Madame Curie,* trans. V. Sheean (Garden City, N.Y.: Garden City Publishing, 1937), 176–77.

9. Blake manuscript reproduced by permission from David Erdman, ed., *The Notebook of William Blake: A Photographic and Typographical Facsimile,* rev. ed. (New York: Readex Books, 1977), N109. The original manuscript of the Blake Notebook, the "Rosetti Manuscript," is in the British Museum, whom we thank for permission to reproduce.

10. Mendeleev draft reproduced by permission from J. W. van Spronsen, *The Periodic System of Chemical Elements* (Amsterdam: Elsevier, 1969), 134.

11. Illustrations modified from M. M. Jones et al., *Chemistry and Society,* 5th ed. (Philadelphia: Saunders, 1987), 28.

12. First published in the *Webster Review* 11, no. 2 (1986): 41.

13. First published in *American Scientist* 78 (1990): 14. The proximity of chemistry to the arts was first stated by Marcellin Berthelot.

14. The Quercetanus sequence is from *Theatrum Chemicum* (1602), quoted in C. G. Jung, *Psychology and Alchemy,* 2d ed., trans. R. F. C. Hull (London: Routledge, 1968), 239. The alchemical passage is ascribed to Hermes in the *Ars Chemica* (1566), quoted also in Jung, *Psychology and Alchemy,* 358–59.

15. I am indebted to A. Truman Schwartz, Macalaster College, for this quotation from R. Watson, *Chemical Essays* (Cambridge, 1781), 1:167.

16. "Ponder Fire" was first published in the *Paris Review* 33, no. 121 (1991): 190.

17. A 10,000-atom model of hemoglobin made by John Mack, Jr.; photographed and copyright by Irving Geis. We are grateful to Mr. Geis for permission to reproduce this image.

18. Hemoglobin subunit structure courtesy of Max Perutz.

19. This essay, under the title "Molecular Beauty III: As Rich as Need Be," was first published in *American Scientist* 77 (1989): 177.

20. First published in *Images* 13, no. 3 (1988): 4.

21. First published in the *Paris Review* 33, no. 121 (1991): 189.

22. An excerpt from the experimental section of an article published in the *Journal of the American Chemical Society* 108 (1986): 4467–76.

23. Figure drawn after one in M. D. Joesten et al., *The World of Chemistry* (Philadelphia: Saunders, 1991), 133.

24. Quotation from H. B. Chipp, *Theories of Modern Art* (Berkeley: University of California Press, 1969), 352.